# SpringerBriefs in Astronomy

**Series Editors**

Wolfgang Hillebrandt
MPI für Astrophysik, Garching, Germany

Michael Inglis
Department of Physical Sciences, SUNY Suffolk County Community College,
Selden, New York, USA

Martin Ratcliffe
Valley Center, Kansas, USA

Steven N. Shore
Dipartimento di Fisica, Enrico Fermi, Università di Pisa, Pisa, Italy

David Weintraub
Department of Physics & Astronomy, Vanderbilt University, Nashville,
Tennessee, USA

SpringerBriefs in Astronomy are a series of slim high-quality publications encompassing the entire spectrum of Astronomy, Astrophysics, Astrophysical Cosmology, Planetary and Space Science, Astrobiology as well as History of Astronomy. Manuscripts for SpringerBriefs in Astronomy will be evaluated by Springer and by members of the Editorial Board. Proposals and other communication should be sent to your Publishing Editors at Springer.

Featuring compact volumes of 50 to 125 pages (approximately 20,000–45,000 words), Briefs are shorter than a conventional book but longer than a journal article. Thus Briefs serve as timely, concise tools for students, researchers, and professionals.

Typical texts for publication might include:

- A snapshot review of the current state of a hot or emerging field
- A concise introduction to core concepts that students must understand in order to make independent contributions
- An extended research report giving more details and discussion than is possible in a conventional journal article
- A manual describing underlying principles and best practices for an experimental technique
- An essay exploring new ideas within astronomy and related areas, or broader topics such as science and society

Briefs allow authors to present their ideas and readers to absorb them with minimal time investment.

Briefs will be published as part of Springer's eBook collection, with millions of readers worldwide. In addition, they will be available, just like other books, for individual print and electronic purchase.

Briefs are characterized by fast, global electronic dissemination, straightforward publishing agreements, easy-to-use manuscript preparation and formatting guidelines, and expedited production schedules.We aim for publication 8–12 weeks after acceptance.

More information about this series at http://www.springer.com/series/10090

Galina O. Ryabova

# Mathematical Modelling
# of Meteoroid Streams

 Springer

Galina O. Ryabova
Research Institute of Applied Mathematics
and Mechanics
Tomsk State University
Tomsk, Russia

ISSN 2191-9100                  ISSN 2191-9119   (electronic)
SpringerBriefs in Astronomy
ISBN 978-3-030-51509-6          ISBN 978-3-030-51510-2   (eBook)
https://doi.org/10.1007/978-3-030-51510-2

This Springer imprint is published by the registered company Springer Nature Switzerland AG.
The registered company address is: Gewerbestrasse 11, 6330 Cham, Switzerland

# Preface

A long time ago, when I was just starting meteoroid stream modelling, working mostly on my own, I had to search through tens of books to solve a simplest problem, e.g. how to generate vectors isotropically. Now I explain these things to students, PhD students, and sometimes to younger colleagues. Gradually, I collected explanatory texts, pictures, etc. The next step—consolidating them in a book—was only logical.

This is a handbook, describing step-by-step computer simulation of meteoroid streams formation and evolution. A detailed derivation of formulae and a large number of figures explaining the subtleties of the method are given. Each theoretical section ends with exercises which allow you to master the material in practice. Most of the examples are based on the Geminid meteoroid stream model, which was being developed by the author in the last 30 years. After the Reader has worked through all the material, (s)he will get a simple model of the stream and the foundation for the construction of his/her own model for another stream.

The book is intended for lower undergraduate, upper-undergraduate (i.e. for bachelor-students and master-students), and PhD students, but also can be used by professionals new to this field. To benefit from this book, students should have some skills in programming, complete introductory courses in meteor astronomy and celestial mechanics, and also know basic concepts of probability theory and statistics.

**Acknowledgments** The writing of this brief was supported by "The Tomsk State University competitiveness improvement programme" under grant (No 8.2.12.2018) and has made use of NASA's Astrophysics Data System Service.

Tomsk, Russian Federation                                             Galina O. Ryabova

# Contents

# Chapter 1
# Introduction

*Mathematical model* is an approximate description of a phenomena or a system, expressed using mathematical symbolism. Here we are going to model a *meteoroid stream*, which is 'a group of meteoroids having similar orbits and a common origin'.[1] Usually under the term 'meteoroid stream model' we understand an ensemble of meteoroids whose ejections from its parent body were modelled and evolution followed.

Meteoroid streams may be produced both by comets and asteroids. In what follows we will consider cometary models, but the only difference between the cometary and asteroidal models is the scheme of meteoroid's release. The general principles are the same.

## 1.1 Mathematical Modelling in General

The process of mathematical modelling of a meteoroid stream can be divided into four stages:

- compiling a physical model of meteoroid stream and cometary decay,
- the choice of mathematical methods,
- comparison with observations and constraining the model,
- revision of the model.

Each of these stages will now be discussed.

---

[1] Recently (April 30, 2017), the International Astronomical Union updated the definitions of terms in meteor astronomy; see (Borovička 2016) or (International Astronomical Union 2017).

© The Author(s), under exclusive license to Springer Nature Switzerland AG 2020
G. O. Ryabova, *Mathematical Modelling of Meteoroid Streams*, SpringerBriefs in Astronomy, https://doi.org/10.1007/978-3-030-51510-2_1

## *1.1.1  Physical Model*

The goal is to know the main structural parameters of a stream, as well as the interrelations between them. These parameters are:

- period of the visibility of the stream on the Earth;
- outbursts (time, location);
- node regression or progression;
- twin showers, if exist;
- flux density profile of the meteor shower;
- profile of the mass distribution index s;
- activity variations along the orbit (or, what is the same, from year to year);
- orbital parameters for meteoroids of different masses;
- radiant (location, motion, configuration);
- grouping of meteoroids in the stream;
- space distribution of meteoroids;
- 'windage' of meteoroids, i.e. $A/m$, where $m$ and $A$ are the mass of the meteoroid and its cross-sectional area, respectively;
- flickering of meteors light-curve.

The parameters are not listed in order of importance. It is regrettable that the uncertainty in some of these parameters is too large even for the most studied streams.

It is very important to know the parent body of a meteoroid stream, because a model constructed on the base of a shower mean orbit (i.e., on the mean orbit of the meteoroids registered at the Earth) cannot give a correct spatial distribution of the meteoroid stream particles.

If the parent body is known, the very first and key parameter to be determined is the age of the meteoroid stream. 'Age' is defined here as the time since the meteoroids were released from the parent body. If the parent comet is active, you might begin modelling from its last return. The stream of a comet, which has been active for a long time, contains meteoroids of different ages. If the parent body is not active, it is necessary to determine the stream age. There are several methods to obtain the stream age, see their description in (Brown and Jones 1998; Ryabova 1999).

To model the decay of the comet (parent body), the following parameters should be known:

- ephemeris of the comet, also in a more distant past, supported by observations;
- dust production rate;
- distribution of ejection velocity vectors in magnitude and direction.

Unfortunately, even for the best observed comets, these parameters are not well determined also.

In the ideal case the list of the output parameters, obtained by modelling, coincides with the list of input parameters, because we verify/constrain a model comparing it with observations.

### 1.1.2  Mathematical Methods

The only choice we may do here is a choice of methods for calculation of the orbital evolution. In a not so remote past, the Gauss–Halphen–Goryachev method was used for integration of orbits, because it is very fast. With this method, however, only secular perturbations of the first order can be calculated, so it cannot be used to address correctly close encounters of meteoroids with planets, or resonance cases. The validity of its application should therefore be carefully verified. For the Geminid meteoroid stream it works perfectly. Moreover, the Geminid stream is so 'unperturbed' that we may even use polynomial approximation (see Sect. 3.3 here) instead of numerical integration.

When the available computing power allows for large scale integration studies, the problem comes to choosing an appropriate integrator and algorithm. Numerical methods of integration of the motion equations are applicable to every kind of motion. But they are time-consuming (even now, when we have super-computers), and have time limitations because of the accumulation of errors. To study long-term evolution a symplectic integrator may be applied (as did, e.g., Brown and Jones 1998). The numerical integration is discussed in Sect. 3.1 here.

It is important to determine exactly what forces are included in the model: which major planets should be taken into account in the calculation of the gravitational perturbations, which radiational perturbations, etc.

### 1.1.3  Constraining the Model and Revision

It is clear that if the model is in accord with observations within the level of errors, it is a good argument in favour of the model reliability. Sometimes we could try to solve an inverse problem, i.e. to constrain initial model parameters in such a way that resulting parameters agree with experimental ones.

Even the most perfect model of its time has limitations determined by the level of experimental data, and, in a more comprehensive sense, by the level of knowledge. A time inevitably comes, when the model should be 'upgraded', or the new model should be constructed. One practical advice: referring to 30-years old results of modelling one should be very careful.

## 1.2   On History of Meteoroid Stream Modelling

Simulation of the meteoroid streams formation and evolution has gone through several stages, which were dictated mainly by computer power. At the first stage, when 'human computers' were used and secular planetary perturbations had to be calculated by hand, efforts of modellers were concentrated on the origin of meteor showers. An excellent example is the prediction of Andromedids 1872 by E. Weiss (see reviews by Ryabova (2006a) or Williams (2011) and references therein).

At the second stage the general dynamics of streams was studied by integrating small quantity of meteoroid orbits (e.g., Williams et al. 1979). Later, when large scale integration studies became possible, McIntosh and Jones (1988) modelled the Comet Halley meteoroid streams, Ryabova (1989) modelled the Geminid stream, Brown and Jones (1998) modelled the Perseid stream, Ryabova (2002) modelled the meteoroid streams of asteroid (1620) Geographos, Vaubaillon et al. (2005) modelled the Leonid meteoroid stream. And now, when we have super-computers in our disposal, the number of papers on meteoroid stream modelling is difficult to calculate.

## 1.3   Several Preliminary Remarks

The following text assumes that the Reader is familiar with the basic concepts of meteor astronomy, celestial mechanics, probability theory, and mathematical statistics. For self-teaching and references I recommend books by Murray and Dermott (2000), Roy (2005), Karttunen et al. (2007), and Wall and Jenkins (2012).

This is an introductory course, so we shall consider simplified models, only *elliptical* orbits and only two of many non-gravitational forces acting on meteoroids (see review by Vaubaillon et al. (2019) and references therein to delve more deeply into the subject).

Most of examples in the chapters below will be given for the Geminid meteoroid stream (in the belief that the stream has cometary origin) and several ones for the Quadrantid meteoroid stream. Exercises are few in number in this book, and I do recommend to complete them all to truly experience the meteoroid stream simulation.

The dust trail theory (Asher 1999, 2000; Sekhar and Asher 2014; Kinsman and Asher 2020) is not considered here. It is aimed at outburst predictions and applicable to compact trails, so either to short time intervals or to resonant meteoroids.

# References

Asher, D.J.: The Leonid meteor storms of 1833 and 1966. Mon. Not. R. Astron. Soc. **307**(4), 919–924 (1999). https://doi.org/10.1046/j.1365-8711.1999.02698.x

Asher, D.J.: Leonid dust trail theories. In: Arlt, R. (ed.) Proceedings of the International Meteor Conference, 18th IMC, Frasso Sabino, pp. 5–21 (2000)

Borovička, J.: About the definition of meteoroid, asteroid, and related terms. WGN J. Int. Meteor Org. **44**, 31–34 (2016)

Brown, P., Jones, J.: Simulation of the formation and evolution of the Perseid meteoroid stream. Icarus **133**, 36–68 (1998). https://doi.org/10.1006/icar.1998.5920

International Astronomical Union: Definitions of terms in meteor astronomy. https://www.iau.org/static/science/scientificbodies/commissions/f1/meteordefinitions_approved.pdf (2017). Accessed 28 Nov 2019

Karttunen, H., Krüger, P., Oja, H., Poutanen, M., Donner, K.J. (eds): Fundamental Astronomy. Springer, Berlin (2007)

Kinsman, J.H., Asher, D.J.: Orbital dynamics of highly probable but rare Orionid outbursts possibly observed by the ancient Maya. Mon. Not. R. Astron. Soc. **493**(1), 551–558 (2020). https://doi.org/10.1093/mnras/staa249

McIntosh, B.A., Jones, J.: The Halley comet meteor stream: numerical modelling of its dynamic evolution. Mon. Not. R. Astron. Soc. **235**, 673–693 (1988). https://doi.org/10.1093/mnras/235.3.673

Murray, C.D., Dermott, S.F.: Solar System Dynamics. Cambridge University Press, Cambridge (2000)

Roy, A.E.: Orbital motion. 4th edition. Institute of Physics Publishing, Bristol (2005)

Ryabova, G.O.: Effect of secular perturbations and the Poynting–Robertson effect on structure of the Geminid meteor stream. Solar Syst. Res. **23**(3), 158–165 (1989)

Ryabova, G.O.: Age of the Geminid meteor stream (review). Solar Syst. Res. **33**, 224–238 (1999)

Ryabova, G.O.: Asteroid 1620 Geographos: II. Associated meteor streams. Solar Syst. Res. **36**(3), 234–247 (2002)

Ryabova, G.O.: Meteoroid streams: mathematical modelling and observations. In: Lazzaro, D., Ferraz Mello, S., Fernandez, J.A. (eds.) Asteroids, Comets, Meteors, IAU Symposium, vol. 229, pp 229–247 (2006). https://doi.org/10.1017/S1743921305006770

Sekhar, A., Asher, D.J.: Resonant behavior of comet Halley and the Orionid stream. Meteorit. Planet. Sci. **49**(1), 52–62 (2014). https://doi.org/10.1111/maps.12117

Vaubaillon, J., Colas, F., Jorda, L.: A new method to predict meteor showers. II. Application to the Leonids. Astron. Astrophys. **439**(2), 761–770 (2005). https://doi.org/10.1051/0004-6361:20042626

Vaubaillon, J., Neslušan, L., Sekhar, A., Rudawska, R., Ryabova, G.: From parent body to meteor shower: the dynamics of meteoroid streams. In: Ryabova, G.O., Asher, D.J., Campbell-Brown, M.D. (eds.) Meteoroids: Sources of Meteors on the Earth and Beyond, chap. 7, pp. 161–186. Cambridge University Press, Cambridge (2019)

Wall, J.V., Jenkins, C.R.: Practical statistics for astronomers. 2nd edition. Cambridge University Press, Cambridge (2012)

Williams, I.P.: The origin and evolution of meteor showers and meteoroid streams. Astron. Geophys. **52**(2), 2.20–2.26 (2011). https://doi.org/10.1111/j.1468-4004.2011.52220.x

Williams, I.P., Murray, C.D., Hughes, D.W.: The long-term orbital evolution of the Quadrantid meteor stream. Mon. Not. R. Astron. Soc. **189**, 483–492 (1979). https://doi.org/10.1093/mnras/189.3.483

# Chapter 2
# Initial Stage: The Model Stream Generation

The method of the stream model construction is quite common. A point where a particle is ejected is chosen on the reference orbit (parent body orbit). Based on a certain assumptions on the ejection scheme (for example, a cometary dust release or a collision, or something else), the velocity vector of the particle is obtained using a generator of pseudo-random numbers, and the particle orbital elements are calculated. Then evolution of these elements is calculated in the given time interval. Repeat this process $N$ times and you will have a model $N$-particle stream. As you see, we use a random sampling method, which is also called Monte Carlo method.

To obtain a qualitative picture of density distribution over a model stream cross-section, 5000 model meteoroids are enough as a rule. However, to calculate the stream activity profile only those meteoroids (or orbits) which intersect the ecliptic plane in the vicinity of the Earth's orbit should be selected. In this case, we should model tens of thousands or even millions of meteoroids. In such long runs you should be careful and not exceed the cycle length of your random-number generator (see the details in Appendix B).

## 2.1 Reference Orbit

Orbital elements of the parent body orbit could be taken either from JPL Small-Body Database Browser (https://ssd.jpl.nasa.gov/sbdb.cgi) or (if it is an asteroid) from the Bowell catalogue (ftp://ftp.lowell.edu.pub/elgb/astrorb.dat). The elements should be integrated to the epoch of the stream generation (see Sect. 3.1).

© The Author(s), under exclusive license to Springer Nature Switzerland AG 2020
G. O. Ryabova, *Mathematical Modelling of Meteoroid Streams*, SpringerBriefs
in Astronomy, https://doi.org/10.1007/978-3-030-51510-2_2

## 2.2   Ejection Point on the Reference Orbit

Let Keplerian elements of the reference orbit be $(a, e, i, \Omega, \omega)$, and the true anomaly of the ejection point is $\upsilon$. We use the system of astronomical constants and units defined by the International Astronomical Union (IAU) (Luzum et al. 2011). The IAU system has units of length (the astronomical unit, au $= 1.49597870700 \times 10^{11}$ m), mass (the mass of the Sun), and time (the day, d). So the speed in this system has dimension [au d$^{-1}$]. In what follows we use the Gauss gravitational constant $k = 0.01720209895$ (au$^3$ d$^{-2}$)$^{1/2}$ and $\mu = k^2$. Luzum et al. (2011) specify: 'For users who need a consistent value for the heliocentric gravitational constant, $k$ should also be used to derive $GM_{\odot}$ through the equation $au^3 k^2 / D^2 = GM_{\odot}$ where $D$ is one day of 86,400 s'. So the heliocentric gravitational constant we use further is $GM_{\odot} = 1.3271244004 \times 10^{20}$ m$^3$ s$^{-2}$.

The first step is to find the position vector of the ejection point $\mathbf{r_c}(x_c, y_c, z_c)$ and the velocity vector of the parent body in this point $\mathbf{V_c}(V_{cx}, V_{cy}, V_{cz})$. The reference system is standard heliocentric ecliptic one with the axes $X$, $Y$, and $Z$.

$$u = \upsilon + \omega,$$
$$p = a(1 - e^2),$$

$$r_c = p/(1 + e \cos \upsilon), \tag{2.1}$$

$$x_c = r_c(\cos u \cos \Omega - \sin u \sin \Omega \cos i),$$
$$y_c = r_c(\cos u \sin \Omega - \sin u \cos \Omega \cos i), \tag{2.2}$$
$$z_c = r_c(\sin u \sin i),$$

$$V_r = \sqrt{\mu/p} \; e \sin \upsilon,$$
$$V_n = \sqrt{\mu/p} \; (1 + e \cos \upsilon),$$

$$V_{cx} = x_c V_r/r_c + (-\sin u \cos \Omega - \cos u \sin \Omega \cos i) V_n,$$
$$V_{cy} = y_c V_r/r_c + (-\sin u \sin \Omega + \cos u \cos \Omega \cos i) V_n, \tag{2.3}$$
$$V_{cz} = z_c V_r/r_c + (\cos u \sin i) V_n.$$

## 2.3  Ejection Velocity Vector

Let $\mathbf{V_e}(V_{ex}, V_{ey}, V_{ez})$ be the ejection velocity vector. Then

$$V_{ex} = V_e \cos T,$$
$$V_{ey} = V_e \sin T \cos \Phi, \qquad (2.4)$$
$$V_{ez} = V_e \sin T \cos \Phi.$$

Here $T$ is the angle between $\mathbf{V_e}$ and the $X$ axis, $T \in [0, \pi]$, $\Phi$ is the angle between $Y$ axis and the projection of $\mathbf{V_e}$ on the $YZ$ plane, $\Phi \in [0, 2\pi]$. Thus, we have to know $T$, $\Phi$, and $V_e$ to calculate $\mathbf{V_e}$.

### 2.3.1  Isotropic Ejection

**Problem**  To generate random vectors $\mathbf{V_e}$, the ends of which are located on a sphere of radius $V_e$, and are evenly distributed on it.

It is known (Korn and Korn 1968, eq. (17.3-13)) that the surface element $d\sigma$ of the sphere can be written as

$$d\sigma = V_e^2 \sin T \, dT \, d\Phi. \qquad (2.5)$$

It follows that in order for the density of points generated on the sphere to be distributed uniformly over its surface, the angle $\Phi$ should be distributed evenly in the interval $[0, 2\pi)$, and the probability density of the angle $T$ should be proportional to $\sin T$.

The fragment of the computer code (in Delphi) is then

```
ksi1:=Random;
ksi2:=Random;
F:=2*Pi*ksi1;
CoT:=1-2*ksi2;
SiT:=Sqrt(1-CoT*CoT);
T:=ArcTan2(SiT,CoT);
```

While the isotropic ejection is a good zero-level approximation, the real comet outgassing is anisotropic. So we need more complicated anisotropic models: the ejection from a nucleus hemisphere facing the Sun, the ejection within a cone (up to jets), and the ejection with intensification to the subsolar point of a nucleus. In this case it is more convenient to use a reference system with an axis directed to the Sun. For jets (not considered here) an axis should be directed along the jet.

### 2.3.2   Ejection in a Cone

Let us introduce a rectangular coordinate system $X'Y'Z'$ so that $X'$ axis was directed towards the Sun, and $Y'$ and $Z'$ were directed arbitrarily, but in such a way that all three axes form a right-handed system. The origin of $X'Y'Z'$ coincides with the origin of $XYZ$. Consider now the ejection velocity vector $\mathbf{V_e}$ in the system $X'Y'Z'$. Then, assuming that $T'$ is the angle between $\mathbf{V_e}$ and the $X'$ axis, and $\Phi'$ is the angle between $Y'$ axis and the projection of $\mathbf{V_e}$ on the $Y'Z'$ plane we may write a set of equations similar to (2.4):

$$V_{ex'} = V_e \cos T',$$
$$V_{ey'} = V_e \sin T' \cos \Phi', \qquad (2.6)$$
$$V_{ez'} = V_e \sin T' \cos \Phi'.$$

Let $\alpha$ be the half-opening of the ejection cone. At a stretch we assume that *within the cone* the ejection is isotropic. It is evident that if $\alpha = \pi$ the ejection is isotropic in a strict sense. So the code below is applicable to both the isotropic ejection and ejection in a cone. Appendix B (Example 1) explains how to generate $T'$ and $\Phi'$ for this case.

The fragment of the code for this model become

```
alpha:=35*pi/180.0;   {for alpha = 35 degrees, e.g.}
ksi1:=Random;
ksi2:=Random;
F:=2*pi*ksi1;                    {Phi'}
CoT:=1-(1-Cos(alpha))*ksi2;      {Cos T'}
SiT:=Sqrt(1-CoT*CoT);            {Sin T'}
T:=ArcTan2(SiT,CoT);             {T'}
```

### 2.3.3   Subsolar Point and the Cosine Ejection

The surface temperature distribution on a rotating atmosphereless body like a cometary nucleus or an asteroid is not uniform. Putting aside physical properties of the body (e.g., albedo, thermal conductivity, etc.) we may reasonably suppose that the temperature depends on the solar radiation flux. So it is the highest in the subsolar point. Measurements in the comet Halley coma have shown that the dust ejections intensity follows the cosine distribution law (see Ryabova 1997, and references therein). Appendix B (Example 2) explains how to generate $T'$ and $\Phi'$ for this case.

The fragment of the code for this model become

```
...
SiT:= Sin(alpha)*sqrt(U);
CoT:= sqrt(1- sqr(SiT));
T:=ArcTan2(SiT,CoT);             {T'}
```

## 2.4 Ejection Speed

The classical model of ejection of dust grains from a cometary nucleus is described in the papers by Whipple (1950, 1951). The second paper contains the famous Whipple formula that is usually referred to. The classical form of Whipple's formula for a spherical particle and a spherical nucleus of a comet is

$$V_\infty = \left( \frac{1}{n \rho_m r_m r^{9/4}} - 0.013 \rho_c R_c \right)^{1/2} \text{cm s}^{-1}. \tag{2.7}$$

Here $1/n$ is a portion of the solar energy expended in sublimation, $\rho$ (g cm$^{-3}$) is the meteoroid density, $r_m$ (cm) is the meteoroid radius, $r$ (au) is the heliocentric distance, $R_c$ (km) and $\rho_c$ (g cm$^{-3}$) are the radius and the density of the cometary nucleus. In the model calculations, the coefficient $1/n$ is usually assumed to be equal to unity. Needless to say that the ejection speed should exceed the escape speed $V_{esc} = \sqrt{2Gm_c/R_c}$, where $m_c$ is the mass of the parent body and $G = 6.67428 \times 10^{-11}$ m$^3$ kg$^{-1}$ s$^{-2}$ is the constant of gravitation (Luzum et al. 2011).

A comprehensive analytical review related to the ejection speed from comets could be found in (Ryabova 2013). In this paper you will find the detailed derivation of the formula (2.7) and a detailed discussion of other existing formulae.

## 2.5 Calculation of the Model Meteoroid Orbit

At this point we have the position vector of the ejection point $\mathbf{r_c}(x_c, y_c, z_c)$, the velocity vector of the parent body in this point $\mathbf{V_c}(V_{cx}, V_{cy}, V_{cz})$, and the ejection velocity vector $\mathbf{V_e}(V_{ex}, V_{ey}, V_{ez})$.

The heliocentric velocity of the $i$th model meteoroid is easily obtained:

$$\mathbf{V_i} = \mathbf{V_c} + \mathbf{V_e}. \tag{2.8}$$

Then the orbital elements $(a_i, e_i, i_i, \Omega_i, \omega_i, \upsilon_i)$ of the $i$th model meteoroid can be computed from $\mathbf{V_i}$ and $\mathbf{r_c}$ using the formulae (A.1–A.10) in Appendix A.

## 2.6 Ejection Along the Orbit

Cometary scenario of the dust ejection implies that the dust production rate increases to perihelion and decreases after its passage. In the first approximation we may suppose that the distribution of meteoroid production is proportional to $r^{-\delta}$, where $r$ is the heliocentric distance. In the review of de Almeida et al. (2007) related to seven comets—possible target comets to space missions—it is indicated that $\delta$ varies from 2.3 till 6.0. Ryabova (2007, 2016) assumed $\delta = 4$ in modelling the Geminid meteoroid stream and the comet Halley meteoroid stream (Ryabova 2003).

How to generate the true anomalies of the ejection points on a reference orbit is explained in Appendix B (Example 3).

## 2.7   A Simple Model: The Unperturbed Stream

Let us consider infant years of a meteoroid stream, i.e. the first and several subsequent orbital periods. It is convenient to do within the two body problem (see Roy 2005, Ch. 4). For simplicity we will assume that all meteoroids are ejected from the comet at a speed $V_{ej}$ at perihelion. For some time the meteoroid swarm stays in the vicinity of the nucleus. But as the periods of meteoroids vary, the swarm gradually spreads around the orbit. Fox et al. (1983) give the approximate loop formation time as

$$P_{max}^2 (P_{max} - P_{min})^{-1}, \tag{2.9}$$

where $P_{max}$ and $P_{min}$ are the maximum and minimum orbital periods of the meteoroids. Evidently a meteoroid has the maximum speed and the maximum period, when it is ejected in the direction of $\mathbf{V_c}$, and the minimum speed and period when it is ejected in the opposite direction.

The period $P$ of the body is given by

$$P = 2\pi \left( \frac{a^3}{\mu} \right)^{1/2}, \tag{2.10}$$

and the semi-major axis can be obtained from the relation

$$V^2 = \mu \left( \frac{2}{r} - \frac{1}{a} \right). \tag{2.11}$$

The last equation is the other form of the energy integral.

Now let us model the loop formation numerically using Kepler's equation

$$M = E - e \sin E, \tag{2.12}$$

where $M$ and $E$ are the mean and the eccentric anomalies, respectively. Tens of methods exist for its solution. The routine kepl1 at the end of this section implements Legendre-based starter and Halley iterator (Odell and Gooding 1986). The algorithm to obtain an unperturbed stream model might look like this:

K1.   [Initialize.] Time $t = t_0$. Ejection of N particles at perihelion. Calculation of
$a_j, e_j, i_j, \Omega_j, \omega_j, \upsilon_j; \quad j \leftarrow 0$.

K2.  [Process the next meteoroid. Loop on $j$. ] Perform steps K3 through K7 for $j = 1, 2, \ldots, N$; then terminate the algorithm.

K3.  [Transition from the true anomaly to the mean anomaly.]  $\upsilon_j \Longrightarrow M_j$.

K4.  [Take time-step.]  $t \leftarrow t_0 + \Delta t$. If $t > t_{end}$ terminate the algorithm.

K5.  [Calculate the new mean anomaly.]  $M_j \leftarrow n\Delta t + M_j$, where $n = 2\pi/P$ is the mean angular velocity (mean motion). Be sure that $n$ is expressed in [rad day$^{-1}$].

K6.  [Transition from the mean anomaly to the true anomaly.]  $M_j \Longrightarrow \upsilon_j$.

K7.  [Output $x_j, y_j, z_j$.]   Calculate the particle position $x_j, y_j, z_j$ for the time $t$. Write onto the output file.

The relation between the true anomaly and the eccentric anomaly (steps K3 and K6) may be written as

$$r \cos \upsilon = a(\cos E - e),$$

$$r \sin \upsilon = a\sqrt{1 - e^2} \sin E. \tag{2.13}$$

The routines v_M and M_v implement transitions between $M$ and $\upsilon$.

```
procedure v_M (a, e, v: Extended; var M: Extended);
   {Transition from the true anomaly v to the mean anomaly M.
    Angles are in radians. Resulting M can be negative.}
var cv,e2,r,sE,cE,EE: Extended;
begin
  cv:=cos(v);
  e2:=sqr(e);
  r:=a*(1-e2)/(1+e*cv);
  sE:=sin(v)*r/(a*sqrt(1-e2));
  cE:=cv*r/a+e;
  EE:=ArcTan2(sE,cE);
  M:=EE-e*sE;
end; {v_M}

procedure M_v(a, e, M: Extended; var v: Extended);
var EE,cE,r,rec_r,sv,cv: Extended;
   {Transition from the mean anomaly M to the true anomaly v.
    Angles are  in radians. Resulting v can be negative.
    Needs function kepl1.}
begin
  EE:=kepl1(M,e);
  cE:=cos(EE);
  r:=a*(1-e*cE);
  rec_r:=1/r;
  sv:=a*sqrt(1-sqr(e))*sin(EE)*rec_r;
  cv:=a*(cE-e)*rec_r;
  v:=ArcTan2(sv,cv);
end; {M_v}
```

And here is a routine for solving Kepler's equation (2.12).

```
function kepl1(M,e: Extended): Extended;
  {Solves Kepler's equation M = E-e sin(E) with Legendre-ba-
  sed starter and Halley iterator. Used as standard in Royal
  Aircraft Establishment (England) under the name EKEPL1.
  Input values: M - mean anomaly; e - eccentricity.
  Output kepl1 is the eccentric anomaly.
  Source: Odell A.W., Gooding R.H. Cel. Mech. 1986,38,4,
  307-336. }
var c,eta,f,fd,fdd,psi,s,xi: Extended;
begin
  c:=e*cos(M);
  s:=e*sin(M);
  psi:=s/sqrt(1-c-c+e*e);
  repeat
    xi:=cos(psi);
    eta:=sin(psi);
    fd:=(1-c*xi)+s*eta;
    fdd:=c*eta+s*xi;
    f:=psi-fdd;
    psi:=psi-f*fd/(fd*fd-0.5*f*fdd);
  until f*f <= 1e-12;
  kepl1:=M+psi;
end; {kepl1}
```

# Exercises

2.1 The orbital elements of the asteroid (3200) Phaethon are: $a = 1.271$ au, $e = 0.890$, $i = 22.164°$, $\Omega = 265.443°$, $\omega = 321.97°$. Generate ejection points around all the orbit so that their true anomalies $v$ were distributed (1) uniformly in $[0, 2\pi)$ and (2) so that the dust production rate was proportional to $r^{-3}$. For these two models make plots of $r$-distribution and explain the result.

2.2 Using the same orbital elements find the period of revolution of the asteroid in days.

2.3 Build a model for 1000 Geminid particles as described in Sect. 2.7. Follow the model over several orbital periods and observe the meteoroids distribution around the orbit using a software for scientific plots (see Chap. 5). Estimate the loop formation time and compare it with the analytic estimation (2.9).

# References

de Almeida, A.A., Sanzovo, G.C., Singh, P.D., Misra, A., Miguel Torres, R., Boice, D.C., Huebner, W.F.: On the relationship between visual magnitudes and gas and dust production rates in target comets to space missions. Adv. Space Res. **39**(3), 432–445 (2007). https://doi.org/10.1016/j.asr.2006.09.039

Fox, K., Williams, I.P., Hughes, D.W.: The rate profile of the Geminid meteor shower. Mon. Not. R. Astron. Soc. **205**, 1155–1169 (1983). https://doi.org/10.1093/mnras/205.4.1155

Korn, G.A., Korn, T.M.: Mathematical Handbook for Scientists and Engineers: Definitions, Theorems, and Formulas for Reference and Review, 2nd enl. and rev. edn. McGraw-Hill, New York (1968)

Luzum, B., Capitaine, N., Fienga, A., Folkner, W., Fukushima, T., Hilton, J., Hohenkerk, C., Krasinsky, G., Petit, G., Pitjeva, E., Soffel, M., Wallace, P.: The IAU 2009 system of astronomical constants: the report of the IAU working group on numerical standards for Fundamental Astronomy. Celest. Mech. Dyn. Astron. **110**(4), 293–304 (2011). https://doi.org/10.1007/s10569-011-9352-4

Odell, A.W., Gooding, R.H.: Procedures for solving Kepler's equation. Celest. Mech. **38**(4), 307–334 (1986). https://doi.org/10.1007/BF01238923

Roy, A.E.: Orbital motion. 4th edition. Institute of Physics Publishing, Bristol (2005)

Ryabova, G.O.: Modeling the ejection of large dust particles from the nucleus of Comet Halley. Solar Syst. Res. **31**(4), 277–288 (1997)

Ryabova, G.O.: The comet Halley meteoroid stream: just one more model. Mon. Not. R. Astron. Soc. **341**, 739–746 (2003). https://doi.org/10.1046/j.1365-8711.2003.06472.x

Ryabova, G.O.: Mathematical modelling of the Geminid meteoroid stream. Mon. Not. R. Astron. Soc. **375**, 1371–1380 (2007). https://doi.org/10.1111/j.1365-2966.2007.11392.x

Ryabova, G.O.: Modeling of meteoroid streams: the velocity of ejection of meteoroids from comets (a review). Solar Syst. Res. **47**, 219–238 (2013). https://doi.org/10.1134/S0038094613030052

Ryabova, G.O.: A preliminary numerical model of the Geminid meteoroid stream. Mon. Not. R. Astron. Soc. **456**, 78–84 (2016). https://doi.org/10.1093/mnras/stv2626

Whipple, F.L.: A comet model. I. The acceleration of Comet Encke. Astrophys. J. **111**, 375–394 (1950). https://doi.org/10.1086/145272

Whipple, F.L.: A comet model. II. Physical relations for comets and meteors. Astrophys. J. **113**, 464. https://doi.org/10.1086/145416

# Chapter 3
# The Model Stream Evolution

Meteoroids are the Solar System bodies, so their motion obeys the laws of celestial mechanics. Here we will consider only gravitational perturbations from planets, the radiation pressure, and the Poynting–Robertson effect (PR-effect). A comprehensive up to date review related to the dynamics of the meteoroid streams and forces acting on meteoroids can be found in (Vaubaillon et al. 2019).

## 3.1 Equations of Motion

If a meteoroid orbits about a massive spherical body (i.e., the Sun), without perturbations by other masses, it moves in a central force field, and its equation of motion is

$$\frac{d^2\mathbf{r}}{dt^2} = -\frac{G(M_0 + m)}{r^2}\frac{\mathbf{r}}{r},\tag{3.1}$$

where $G = 6.67428 \times 10^{-11}\,\mathrm{m^3\ kg^{-1}\ s^{-2}}$ is the constant of gravitation (Luzum et al. 2011), $M_0$ is the mass of the Sun, $m$ is the meteoroid mass, and $\mathbf{r}$ is the position vector of the meteoroid. Taking into account that $m \ll M_0$, $GM_0 = k^2$ and $\mu = k^2$ we obtain

$$\frac{d^2\mathbf{r}}{dt^2} = -\frac{\mu}{r^2}\frac{\mathbf{r}}{r}.\tag{3.2}$$

© The Author(s), under exclusive license to Springer Nature Switzerland AG 2020
G. O. Ryabova, *Mathematical Modelling of Meteoroid Streams*, SpringerBriefs in Astronomy, https://doi.org/10.1007/978-3-030-51510-2_3

Let us add several planets to this force model. Then the meteoroid equation of motion become

$$\frac{d^2\mathbf{r}}{dt^2} = -\frac{\mu}{r^2}\frac{\mathbf{r}}{r} + G\sum_{j=1}^{n} m_j \left( \frac{\mathbf{r}_j - \mathbf{r}}{|\mathbf{r}_j - \mathbf{r}|^3} - \frac{\mathbf{r}_j}{r_j^3} \right), \quad (3.3)$$

where $m_j$ is the mass of the $j$th perturbing body, $\mathbf{r}_j$ is the position vector of this body. Solving this second-order linear differential equation we obtain $\mathbf{r}(t)$. Equatorial coordinates of all the planets and the Moon might be taken from the JPL Planetary Development Ephemeris (https://ssd.jpl.nasa.gov/?planet_eph_export).

Theory and practice of the numerical integration are beyond the scope of this book. There are open software sources, e.g. widely known software package MERCURY (Chambers 1999).[1] An analytical overview of six common integration algorithms may be found in Eggl and Dvorak (2010).

In the majority of cases we have to use numerical integration, because the meteoroids orbits suffer appreciable perturbations. However, there is at least one case, when we may use polynomial approximations instead. This case of the Geminid meteoroid stream will be considered in the Sect. 3.3.

## 3.2   Reasonable Time of the Integration

Any system has its limit of predictability. In celestial mechanics Lyapunov time—the characteristic timescale on which a dynamical system is chaotic—is used to determine the reasonable length of integration time. For example, for the asteroid (3200) Phaethon the chaos onset happens after about 3 kyr of the backward integration (Ryabova et al. 2019), and for the asteroid (196256) 2003EH1 Lyapunov time is $\approx 80$ years (Abedin et al. 2015).

However, we need not go into these matters too deep. A more simple way exists to evaluate when the chaos begins to develop. This technique is described in (Ryabova et al. 2019), where it was used to study dynamics of three asteroids, including (3200) Phaethon.

To follow the backward evolution of the asteroids 200 clones were created for the asteroid, using the method described in details by Williams et al. (2004, section 5). The equations of motion for the asteroids and its clones were numerically integrated 5000 years back. It is apparent (Fig. 3.1) that the divergence of the clones begins at about 3.5–4 kyr ago, but the dispersion in the orbital elements is small, and during the first 3 kyr the nominal orbit and the clones are almost indistinguishable. It is worth adding that until the year zero the integration accuracy was very good and quite acceptable. However, thereafter, the accuracy decreased swiftly in time. The reason for this loss of accuracy is probably due to the onset of chaos.

---

[1] See the entry 1201.008 in the Astrophysics Code Library ASCL.net.

**Fig. 3.1** The dispersion of the semimajor axis $\Delta a$ of the clones relatively the nominal orbit of (3200) Phaethon. Time is counted from the epoch 2000.0. This figure is a part of (Ryabova et al. 2019, fig. 2), where $\Delta e$ and $\Delta i$ are shown also. Behavior of $\Delta e$ and $\Delta i$ is similar

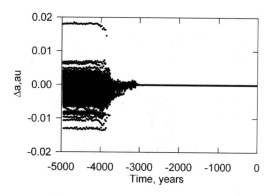

**Fig. 3.2** The asteroid Phaethon (solid line) and the mean Geminid orbits (dotted line) in projection to the ecliptic plane. The grey ring designates the asteroid belt. Modified after (Ryabova et al. 2019, fig. 1)

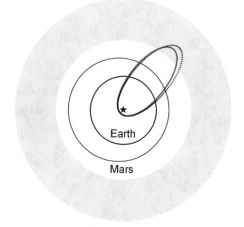

To summarize. Be very careful integrating beyond the time when the chaos begins to develop. The results should be treated with caution, and mostly statistically.

## 3.3 Half-Analytical Approach

The orbits of the Geminid meteoroid stream as well as that of the asteroid (3200) Phaethon (the Geminid's parent body) are located far inside Jupiter's orbit (Fig. 3.2). The asteroid has no close encounters significantly affecting its orbit, during the last 10,000 years at least. That is why its semimajor axis is practically constant, and the changes in the asteroid's orbital elements are smooth (Fig. 3.3).

Considering that the same is true for the most part of the Geminid meteoroids, the changes in orbital elements may be approximated by a set of nested polynomials of the form

$$b(t) = b_0 + \sum_{j=1}^{n} \sum_{k=0}^{m} b_{jk} a_0^k t^j, \qquad (3.4)$$

**Fig. 3.3** Orbital elements of asteroid (3200) Phaethon as a function of time. Numerical integration by the Everhart method (solid line), the Halphen–Goryachev method (crosses) when six planets (Venus–Uranus), and only Jupiter (dashed line) were taken into account. The top graph here and in Fig. 3.4 has no crosses for *a*, because *a* is constant when using Gauss-type methods. After (Ryabova 2007, fig. 2)

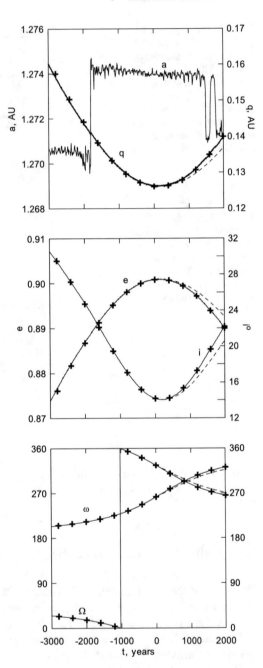

where $b$ is one of the Keplerian elements $(a, e, i, \Omega, \omega)$, $t$ is the time from the initial moment (epoch of the stream generation), $b_0$ is the initial value for an element (at $t = 0$); $n$ and $m$ are determined individually for each element, but as a rule $m = 3$, $n = 3, 4$. Nested polynomials are polynomials whose coefficients are also polynomials. For example, the eccentricity for a model meteoroid is calculated by the following way.

$$e = e_0 + e_1 t^1 + e_2 t^2 + e_3 t^3 + e_4 t^4,$$

where

$$e_j = e_{j0} a_0^0 + e_{j1} a_0^1 + e_{j2} a_0^2 + e_{j3} a_0^3.$$

The polynomial coefficients and parameters of the model are given in Appendix C.

The applicability of this method is very narrow: at the moment only the Geminid meteoroid stream. (Another possible candidate for future study could be the Daytime Arietid meteoroid stream.) So there is no need to explain the details here. How to calculate the polynomial coefficients was described in (Ryabova 2007). A brief summary: the evolution of 20–30 orbits of a model stream is calculated by a numerical method and then the least-square fit to functions $b(t)$ obtained from the numerical integration is used. This method is very fast. Calculation of orbital evolution for millions of orbits on the time-span of two thousand years takes a couple of minutes on a usual desktop computer. So it is very suitable for educational purposes. The first version of the polynomial coefficients was published in (Ryabova 1989). Later Ryabova (2007) used two sets of nested polynomials instead of a single one. The first was used for the main part of the stream $[a_{min}, a_{lim}]$, and the second for the 'tail' $[a_{lim}, a_{max}]$.

The latter requires some explanation. Look closely at the model Geminid cross-section (Fig. 5.2 here or better Ryabova 2007, fig. 5). The dispersion in space of the model stream is extremely anisotropic. The stream expands significantly, but not because of change in width. About 90% of the model meteoroids are concentrated around the mean orbit, while the remaining 10% produces a rarefied long 'tail'. So it was decided to make approximations separately for these two parts.

A good alternative is using a Gauss-type method, for example, Halphen–Goryachev method (Goryachev 1937; Sukhotin 1981). This is also a fast method, however not so fast as the nested polynomials. Calculations of the asteroid (3200) Phaethon orbital evolution made by the Halphen–Goryachev method, where perturbations from six planets (Venus–Uranus) were taken into account, visibly coincide with results of the Everhart method (Fig. 3.3). Both Halphen–Goryachev and nested polynomials methods are applicable for modelling the Geminid stream. It is not so for the Quadrantid stream. Aphelion of asteroid (196256) 2003EH1 (parent body of the Quadrantid meteoroid stream) is located near orbit of Jupiter. The Quadrantid meteoroids have close approaches with Mercury, Venus, Earth, and Jupiter, so only a precise numerical integration method may be used for the stream

**Fig. 3.4** Orbital elements of
asteroid (196256) 2003EH1
as a function of time.
Numerical integration by the
Everhart method (solid line)
and the Halphen–Goryachev
method (crosses) when six
planets (Venus–Uranus) were
taken into account. Modified
after (Ryabova 2006, fig. 1)

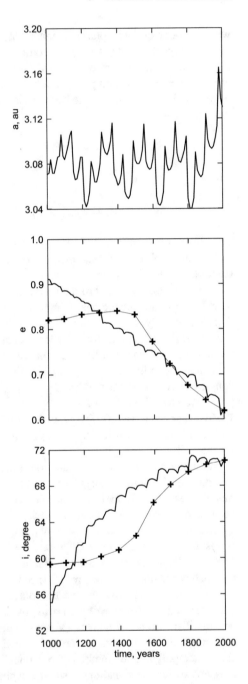

modelling (Fig. 3.4). So the first step in the study of a model stream evolution should
be the careful analysis of the factors influencing the dynamical evolution of the
stream's parent body.

## 3.4 How to Take into Account the Radiation Pressure and Poynting–Robertson Drag Forces

The equation of motion for a particle of mass $m$ and geometrical cross-section $A$, moving with velocity $V$ through a radiation field of energy flux density $S$, is (to terms of order $V/c$)

$$\frac{d\mathbf{V}}{dt} = \left(\frac{SA}{cm}\right) Q_{pr} \left[\left(1 - \frac{V_r}{c}\right) \hat{\mathbf{S}} - \frac{\mathbf{V}}{c}\right], \tag{3.5}$$

according to (Burns et al. 1979, eq. (5)). Here $\hat{\mathbf{S}}$ is a unit vector in the direction of the incident radiation, $Q_{pr}$ is the coefficient of Mie scattering (= 1 for our size range, i.e. hundred-micron and more), $c$ is the speed of light, $V_r$ is the meteoroid's radial velocity, $S$ is the solar energy flux density at distance $r$. We can write also $S = S_0(r_0/r)^2$, where $S_0$ is the solar constant and $r_0 = 1$ au in centimetres. $S_0 = 1.367 \times 10^6 \, \mathrm{g\,s^{-3}}$ according to (IERS Standards 1992), or $S_0 = 1.3611 \times 10^6 \pm 0.0005 \, \mathrm{g\,s^{-3}}$ according to (Gueymard 2018).

Following Burns et al. (1979) we call the constant radial term of (3.5) the *radiation pressure* and the velocity dependent terms the *Poynting–Robertson drag*.

### 3.4.1 Radiation Pressure Force

By Eq. (3.2) the gravitational force of the Sun upon a meteoroid is

$$F_{gr} = \frac{\mu m}{r^2}. \tag{3.6}$$

By Eq. (3.5) the radiation pressure force upon the meteoroid is

$$F_r = \frac{SA}{c} Q_{pr}. \tag{3.7}$$

Their ratio has well-established designation $\beta$:

$$\beta = \frac{F_r}{F_{gr}} = \left(\frac{SA}{cm}\right) Q_{pr} \frac{r^2}{\mu} = 7.7 \times 10^{-5} \frac{A}{m}, \tag{3.8}$$

where the numerical coefficient is valid for $A$ in [cm$^2$] and $m$ in [g]. Hence an addition to the right-hand side of (3.3) is

$$\frac{d^2\mathbf{r}}{dt^2} = -\frac{\mu\beta}{r^2} \frac{\mathbf{r}}{r}, \tag{3.9}$$

and all we have to do is to reduce the mass of the Sun by replacing $\mu$ by $\mu'$

$$\mu' = (1 - \beta)\mu. \qquad (3.10)$$

### 3.4.2   Poynting–Robertson Effect (Photon Component)

The addition to the right-hand side of the Eq. (3.3) is obvious

$$\frac{d^2\mathbf{r}}{dt^2} = -\left(\frac{SA}{cm}\right)\left[\frac{V_r\hat{\mathbf{S}}}{c} + \frac{\mathbf{V}}{c}\right], \qquad (3.11)$$

where $\hat{\mathbf{S}} = \mathbf{r}/r$ and

$$V_r = V \times \cos\angle(\mathbf{V}, \mathbf{r}) = \frac{xV_x + yV_y + zV_z}{r}.$$

Here (and further) $Q_{pr} = 1$ is omitted.

### 3.4.3   Poynting–Robertson Effect (Corpuscular Component)

Ryabova (2005) adapted the theory to use it in meteoroid dynamics modelling. As a rule, the influence of the corpuscular PR-analogue was (and is) taken into account by introducing the proper coefficient equal to the ratio between the corpuscular and radiation drags. The estimates of the different authors give close values of the order of 0.2. However, estimation by Ryabova (2005) for the Geminid meteoroid stream shows that the ratios of corpuscular to radiation drags are 0.4–0.7 for particles of size $> 10\,\mu$m, i.e. the effect is much stronger than was assumed before.

$$\mathbf{F_w} = f_w\frac{\mathbf{U}}{U}. \qquad (3.12)$$

In this expression $\mathbf{U} = \mathbf{W} - \mathbf{V}$, where $\mathbf{W}$ is the average velocity of the solar wind flow and $\mathbf{V}$ is the meteoroid velocity. The approximate expression for $f_w$ is

$$f_w \approx 1.626 \times 10^{-23}\psi\left(\frac{r_0}{r}\right)^2 U^2 A. \qquad (3.13)$$

Here $A$ is the area of the cross-section of the meteoroid. Mukai and Yamamoto (1982) calculated $\psi$ in the range $0.1 < r < 5$ au for the average parameters of the solar wind flow and for three grain materials: water ice, magnetite, and obsidian. The value $\psi$ proved to be nearly constant for these materials, and the following values can be used for estimations: 1.6 (water ice), 1.4 (magnetite), and 1.1 (obsidian). All values in (3.13) are given in CGS units.

Value of $U$ changes along the meteoroid's orbit, but for our estimation it is quite enough to use $U$ averaged over the orbit.

$$\langle U^2 \rangle = \frac{1}{PH} \int_0^\pi U^2 r^2 \, dv = \pi \mu a^{3/2} \left(1 - 2p^2\right) p^{-5/2} - 4Wae(\mu\pi)^{1/2}.$$

(3.14)

Here $P$ is the orbital period; $p = a(1 - e^2)$ is the semilatus rectum for ellipse; $H = [\mu p]^{1/2}$ is the meteoroid angular momentum per unit mass; $v$ is the true anomaly; $W$ is the solar wind speed, and for estimations we may use the approximate mean value of $400 \, \text{km s}^{-1}$.

Solar wind blows not quite radially away from the Sun, and the expression for its velocity can be written in the form

$$\mathbf{W} = W \cos\phi \hat{\mathbf{S}} \pm \sin\phi \hat{\boldsymbol{\theta}},$$

where $\hat{\mathbf{S}}$ is a unit vector in the direction of the incident radiation (as above), and the term of $\sin\phi$ corresponds to the tangential component of the solar wind. Considering that $\phi \approx 1.5°$, we may neglect the second term and take $\mathbf{W} = W\hat{\mathbf{S}}$. Hence the addition to the right-hand side of the Eq. (3.3) is

$$\frac{d^2\mathbf{r}}{dt^2} = \frac{f_w}{mU}(\mathbf{W} - \mathbf{V}).$$

(3.15)

> **Important**
Inserting your additions into the code of your integrator, attend to the optimization of arithmetical operations. The more operations, the greater the rounding error.

### 3.4.4   Verification

There is no direct way to test the correctness of your code after you made all additions to your integrator. However, there are two indirect ways. The first one is obvious: to find somebody's similar independently written program and compare the results. The second way is to compare the variations in the meteoroid orbital semimajor axes and eccentricities obtained numerically with secular variations obtained analytically.

The first set of equations gives the averaged changes in $a$ and $e$ due to PR-effect (Burns et al. 1979, eqs. (47)–(49)). There is no normal force component, so there is no change in the inclination $i$.

$$\left\langle \frac{da}{dt} \right\rangle_{ph} = -4.8 \times 10^{-8} \frac{A}{m} \frac{2 + 3e^2}{a(1 - e^2)^{3/2}} \quad \text{au year}^{-1} ,$$

$$\left\langle \frac{de}{dt} \right\rangle_{ph} = -4.8 \times 10^{-8} \frac{A}{m} \frac{5e}{2a^2(1 - e^2)^{1/2}} \quad \text{year}^{-1} . \tag{3.16}$$

The second set of equations gives the same, but for the corpuscular part of PR-drag (Ryabova 2005, eq. (3.5)).

$$\left\langle \frac{da}{dt} \right\rangle_{w} = -5.15 \times 10^{-16} \psi \hat{U} \frac{A}{m} \frac{2 + 2e^2}{a(1 - e^2)^{3/2}} \quad \text{au year}^{-1} ,$$

$$\left\langle \frac{de}{dt} \right\rangle_{w} = -5.15 \times 10^{-16} \psi \hat{U} \frac{A}{m} \frac{2e}{a^2(1 - e^2)^{1/2}} \quad \text{year}^{-1} . \tag{3.17}$$

In the two equations above $a$ is in [au], $\hat{U}$, $A$ and $m$ are in CGS units [g, cm, s].

> **Important**
Check and double check units in the formulae you are going to use.

### 3.4.5   On Age of a Meteoroid Stream

From Eqs. (3.16) and (3.17) it is seen that a systematic decrease in the semimajor axis and the eccentricity exists, and that its rate depends on the meteoroid mass.

Indeed, several researches found such correlation for the observed Geminids (see Ryabova (1999) and references therein). Having functions $a(m)$ and/or $e(m)$, or just the differences in the semimajor axis between large and small meteoroids, we can estimate the age from (3.16) and (3.17).

The weak points in this method are evident. First, the action of other forces can change the rate of this decrease (Klačka and Kocifaj 2008, e.g.). Second, the real shape of meteoroids most likely deviates from the ideal spherical, therefore, $A/m$ takes larger values (Ryabova 1999). Third, the observed shower may include meteoroids of various age. And the last, the accuracy of the determination of $a(m)$ and $e(m)$ is still not high. Nevertheless, it is advisable to conduct research in this direction for the stream of your choice.

## Exercises

3.1 Repeat exercise 2.3 taking into consideration the mass reduction of the Sun—see Eq. (3.10). Let meteoroids are spherical, their density is $1\,\mathrm{g\,cm^{-3}}$, and their mass is $3 \times 10^{-5}\,\mathrm{g}$. Is there any noticeable difference with the results of the exercise 2.3?

3.2 Belkovich (1986) derived the following relationships for the Geminids

$$a = 1.4110 - 0.0146\,\mathrm{m}^{-1/3}\,, \qquad e = 0.9000 - 0.0009\,\mathrm{m}^{-1/3}\,.$$

Find the stream age, taking the density of meteoroids to be $1\,\mathrm{g\,cm^{-3}}$.

## References

Abedin, A., Spurný, P., Wiegert, P., Pokorný, P., Borovička, J., Brown, P.: On the age and formation mechanism of the core of the Quadrantid meteoroid stream. Icarus **261**, 100–117 (2015). https://doi.org/10.1016/j.icarus.2015.08.016

Belkovich, O.I.: The spatial structure of the Geminids. Astronomicheskij Vestnik **20**, 142–151 (1986)

Burns, J.A., Lamy, P.L., Soter, S.: Radiation forces on small particles in the Solar System. Icarus **40**(1), 1–48 (1979). https://doi.org/10.1016/0019-1035(79)90050-2

Chambers, J.E.: A hybrid symplectic integrator that permits close encounters between massive bodies. Mon. Not. R. Astron. Soc. **304**(4), 793–799 (1999). https://doi.org/10.1046/j.1365-8711.1999.02379.x

Eggl, S., Dvorak, R.: An introduction to common numerical integration codes used in dynamical astronomy. In: Souchay, J., Dvorak, R. (eds.) Dynamics of Small Solar System Bodies and Exoplanets. Lecture Notes in Physics, vol. 790, pp. 431–480. Springer, Berlin (2010)

Goryachev, N.N.: Halphen's Method for Calculating Secular Perturbations of Planets and Its Application to Ceres (in Russian). Tomsk State University, Tomsk (1937)

Gueymard, C.A.: A reevaluation of the solar constant based on a 42-year total solar irradiance time series and a reconciliation of spaceborne observations. Solar Energy **168**, 2–9 (2018). https://doi.org/10.1016/j.solener.2018.04.001

Klačka, J., Kocifaj, M.: Times of inspiralling for interplanetary dust grains. Mon. Not. R. Astron. Soc. **390**(4), 1491–1495 (2008). https://doi.org/10.1111/j.1365-2966.2008.13801.x

Luzum, B., Capitaine, N., Fienga, A., Folkner, W., Fukushima, T., Hilton, J., Hohenkerk, C., Krasinsky, G., Petit, G., Pitjeva, E., Soffel, M., Wallace, P.: The IAU 2009 system of astronomical constants: the report of the IAU working group on numerical standards for Fundamental Astronomy. Celest. Mech. Dyn. Astron. **110**(4), 293–304 (2011). https://doi.org/10.1007/s10569-011-9352-4

Mukai, T., Yamamoto, T.: Solar wind pressure on interplanetary dust. Astron. Astrophys. **107**(1), 97–100 (1982)

Ryabova, G.O.: Effect of secular perturbations and the Poynting–Robertson effect on structure of the Geminid meteor stream. Solar Syst. Res. **23**(3), 158–165 (1989)

Ryabova, G.O.: Age of the Geminid meteor stream (review). Solar Syst. Res. **33**, 224–238 (1999)

Ryabova, G.O.: On the dynamical consequences of the Poynting-Robertson drag caused by solar wind. In: Knežević, Z., Milani, A. (eds.) Proceedings of IAU Colloq. 197: Dynamics of Populations of Planetary Systems, pp. 411–414. Cambridge University Press, Cambridge (2005). https://doi.org/10.1017/S1743921304008920

Ryabova, G.O.: Meteoroid streams: mathematical modeling and observations. In: Verbeeck, C., Wislez, J.M. (eds.) Proceedings of the Radio Meteor School, pp. 67–76 (2006)

Ryabova, G.O.: Mathematical modelling of the Geminid meteoroid stream. Mon. Not. R. Astron. Soc. **375**, 1371–1380 (2007). https://doi.org/10.1111/j.1365-2966.2007.11392.x

Ryabova, G.O., Avdyushev, V.A., Williams, I.P.: Asteroid (3200) Phaethon and the Geminid meteoroid stream complex. Mon. Not. R. Astron. Soc. **485**(3), 3378–3385 (2019). https://doi.org/10.1093/mnras/stz658

Sukhotin, A.A.: Algorithm of the method Gauss–Halphen–Goryachev in Lagrangian variables and its machine realization. Astronomiya i geodeziya **9**, 67–73 (1981)

Vaubaillon, J., Neslušan, L., Sekhar, A., Rudawska, R., Ryabova, G.: From parent body to meteor shower: the dynamics of meteoroid streams. In: Ryabova, G.O., Asher, D.J., Campbell-Brown, M.D. (eds.) Meteoroids: Sources of Meteors on the Earth and Beyond, chap. 7, pp. 161–186. Cambridge University Press, Cambridge (2019)

Williams, I.P., Ryabova, G.O., Baturin, A.P., Chernitsov, A.M.: The parent of the Quadrantid meteoroid stream and asteroid 2003 EH1. Mon. Not. R. Astron. Soc. **355**(4), 1171–1181 (2004). https://doi.org/10.1111/j.1365-2966.2004.08401.x

# Chapter 4
# The End Stage: The Model Stream Now

At this stage of modelling, after calculation of the stream evolution we have a model stream of $N$ orbits, where $N$ can be from, say, 100 orbits to millions of orbits. The stream may consist of sub-streams having the different ages.

All structural parameters used to compare the model and the real meteoroid streams are listed in Sect. 1.1. Three of them (at least), namely flux density profile of the meteor shower, mass distribution profile, and radiant configuration, need some explanations in the style 'how to do'.

## 4.1 The Activity Profile of the Model Meteor Shower

We define the activity profile (curve) of a meteor shower as the number of particles registered at the Earth as a function of time (or solar longitude $\lambda_\odot$). Certainly, if we consider as 'registered' only meteoroids really intersecting with the Earth, the number of modelled meteoroids should be comparable with the real number of meteoroids in the stream. In practice *meteoroids* having nodes on the distance $\Delta$ within the Earth's orbit (or the Earth itself) are referred to as Earth-intersecting. Value of $\Delta$ depends of the stream and the aim of modelling. This approach is good, when the age of the stream is relatively small. Otherwise we may not be sure in the calculated meteoroid position (see Sect. 3.2).

When the stream is old and their meteoroids are dispersed around the entire orbit we may count not particles, but *orbits*, which nodes approach the Earth orbit on the distance $< \Delta$. Ryabova (2007, 3.2) has shown that for the Geminid meteoroid stream this approach is valid.

The *incident flux density* (or flux density, or just flux) of a meteor shower is the number of meteoroids per unit time across unit area normal to the radiant direction. In observations, the cumulative flux density $Q(m_0) = Q(m > m_0)$ is determined, i.e. flux for particles having masses larger than some minimal mass $m_0$. In the model

© The Author(s), under exclusive license to Springer Nature Switzerland AG 2020
G. O. Ryabova, *Mathematical Modelling of Meteoroid Streams*, SpringerBriefs in Astronomy, https://doi.org/10.1007/978-3-030-51510-2_4

we consider the differential flux $q(m)$, i.e. flux for particles with a definite mass, say, $m_3$ or $m_4$. For a model the activity curve is equivalent to number of meteoroids in a unit of time, and it may be considered as the flux curve. However, it should be emphasized that it is *not* proportional to rate profiles [or Zenith Hourly Rate (ZHR) profiles] obtained from observations, because the rate profiles are not free from observational selection. Therefore it is very important that for a model adjusting an incident flux density profile, where observational selection factors are eliminated, was used.

To obtain the activity curve of the model shower the nodes should be projected on the Earth orbit. The most simple (and seems to be logical) way of projection is the normal projection, practically coinciding with the projection along the position vector (line $a$ in Fig. 4.1). If we use the normal projection, a model activity profile will be distorted (Fig. 4.1). And it will be the more distorted the wider is the registration band. More details about this technique you may find in Ryabova (2016). I can only add that the best method to understand what band is suitable for your model is experimenting.

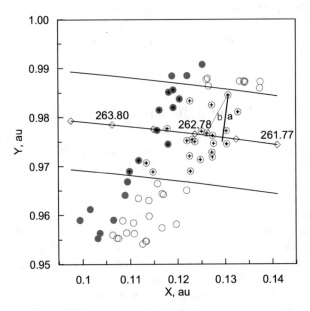

**Fig. 4.1** Geminid model cross-section in the ecliptic plane at the descending node for a stream of 1000 orbits of particles with masses 0.02 g. The reference system is standard heliocentric ecliptic one. Ticks (open diamond) on the Earth orbit are placed every half a day, and labels mark 12 h of a day (J2000) in solar longitude. Nodes of the orbits of meteoroids ejected (open circle) before perihelion passage, and (filled circle) after perihelion passage are shown (will be explained in Sect. 5.2). Crosses mark the nodes of 'registered' 30 meteoroids in the band along the Earth orbit having width $2\Delta = 0.02$ au. Lines $a$ and $b$ show directions of the nodes projection on the Earth orbit: $a$—the normal projection, $b$—along the section (also see the text). After Ryabova (2016, fig. 2)

An example of several model activity profiles see in Fig. 4.2 on the small panels $A$–$E$. Here $\Delta = 4.57 \times 10^{-5}$ au. We could use the narrow band, because the model streams (consisted of meteoroids with masses $m_3$ and $m_4$)[1] had three millions of meteoroids each ensuring statistically rich model showers. For instance, the model meteor shower $m_3$ (i.e., the profile $A$ along the Earth orbit; the profiles $B$–$E$ are explained in Sect. 5.2) has 14.5 thousand of registered meteoroids. An example of statistically pure profiles is presented on Fig. 4.3. This figure demonstrates how the activity profile for the stream $m_2$ changes with the band narrowing. Shape of the curve holds, but its width decreases on about 0.3°. Two alternatives exist to deal with the problem: (1) to obtain an optimal width of the band $\Delta$ experimentally, i.e. to find a reasonable balance between statistical stability of the activity profile and the distortion because of too wide $\Delta$; the more meteoroids in the model, the better is the profile, and the large is the calculation time; (2) to project nodes on the Earth orbit along the general path of the node's secular motion in the ecliptic, i.e. along the stream section (line $b$ in Fig. 4.1). The second technique being more logical than the normal projection is much more complicated in realization.

## 4.2 Mass Distribution in the Model Shower

It is known that mass distribution in a meteoroid stream obeys a power law, so the differential flux can be written as

$$q(m) = const \cdot m^{-s},$$

where $s$ is parameter of the mass distribution. Having model flux for two meteoroid masses we can calculate $s$ from the following

$$s = s_0 + \lg \frac{q(m_4)}{q(m_3)},$$

where $s_0$ is initial value of the parameter, $q(m_3)$ and $q(m_4)$ are differential fluxes obtained for model streams with equal number of model meteoroids. Model $s$-profiles for the Geminid stream are shown in Fig. 4.2.

We know nothing about the initial, i.e. just after ejection from a cometary nucleus, mass distribution. So for the comparison with observations we may use only the shape of the $s$-profile.

---

[1] As almost all examples are taken from the Geminid models (Ryabova 2007, 2016), their initial parameters are given in Appendix C.1.

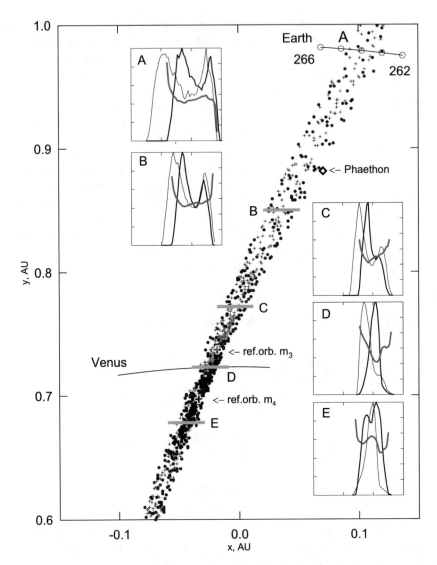

**Fig. 4.2** Geminid model cross-section in the ecliptic plane for orbits of particles with masses $m_3 = 2.14 \times 10^{-3}$ g (plus symbol) and $m_4 = 2.14 \times 10^{-4}$ g (filled circle). Symbol $A$ designates the Earth's orbit in the interval 262–266° in solar longitudes. Other sections are designated by $B - E$. In the small panels, designated by $A - E$, activity profiles, i.e. flux density variations along the Earth's orbit, for particle masses $m_3$ (thick line) and $m_4$ (thin line), and a profile for mass index $s$ (thickest line) are shown. The distance between the tick marks on the abscissa-axes of small panels is equal to $1°$. The profiles are calculated along the corresponding sections. The small panel designated by 'Stream $m_3$' demonstrates pre- and post-perihelion layers (explanation see in Sect. 5.2) in the cross-section of the model stream $m_3$ at the descending node of its mean orbit, designated 'ref. orb. $m_3$'. The plane of the plot is normal to velocity vector of the orbit in the node. The abscissa-axis is directed away from the Sun, the scales on both axes are in au. Modified after Ryabova (2008, fig. 1)

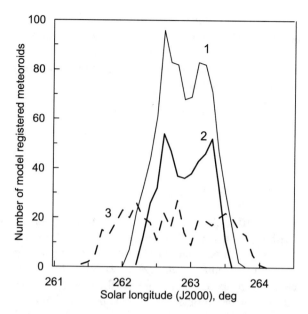

**Fig. 4.3** Model activity curves for model streams in 30 thousand particles: the number of registered particles versus solar longitude. (1) The mass of particle is $m_2$, $\Delta = 0.01$ au (895 nodes). (2) The mass of particle is $m_2$, $\Delta = 0.005$ au (469 nodes). (3) The mass of particle is $m_4$, $\Delta = 0.01$ au (411 nodes). Rate ($N$) is in meteors per $0.1°$ in solar longitude. After Ryabova (2016, fig. 3)

## 4.3   Model Radiants

*Radiant* is the point where the backward projection of the meteor trajectory intersects the celestial sphere. More generally, the point in the sky where meteors from a specific shower seem to come from.[2] It is customary to use the geocentric equatorial coordinates (right ascension $\alpha$ and declination $\delta$) for the radiant point.

When a meteor was observed at two or more stations it is possible to obtain the spatial trajectory of the meteor, and hence its apparent radiant $\alpha_a$, $\delta_a$ and the apparent velocity $\mathbf{V}_a$. Correcting for the deceleration in the Earth's atmosphere, the Earth's rotation, and the Earth's gravitational attraction, we find the true geocentric velocity of the meteoroid, and therefore the true radiant point $\alpha$, $\delta$. The reduction method is not simple, it has to rely on physical models, so the resulting data have errors. For example, speed of high-precision Quadrantids (Abedin et al. 2015) has the uncertainty up to $300\,\mathrm{m\,s^{-1}}$, and for the Geminids (Hajduková et al. 2017) up to $210\,\mathrm{m\,s^{-1}}$.

---

[2]According to the International Meteor Organization glossary: https://www.imo.net/resources/glossary/.

In modelling the situation is much easier. If a meteoroid approaches the Earth at a distance less than the registration distance $\Delta$, we began to keep track of its position. The aim is to find the place where the distance meteoroid–Earth is minimal, and calculate the radiant for this position. In the following expressions prime denotes belonging to the heliocentric *equatorial* system.

$$\mathbf{V}'_G = \mathbf{V}' - \mathbf{V}'_E \,,$$
$$\sin \alpha = V'_{Ey} - V'_y \,,$$
$$\cos \alpha = V'_{Ex} - V'_x \,, \tag{4.1}$$
$$\sin \delta = (V'_{Ez} - V'_z)/V'_G \,.$$

Here $\mathbf{V}'_G$ is the geocentric velocity of a meteoroid, $\mathbf{V}'$ is its heliocentric velocity, and $\mathbf{V}'_E$ is the Earth's velocity. All other notations are self-explanatory.

Observed meteoroids impact the Earth, but the model meteoroids do that infrequently. So the model radiant points (Fig. 4.4) have a bias, and the farther the orbit of a meteoroid from the Earth, the greater the bias. I do recommend you to study the bias for your model. It might be worthwhile to reduce $\Delta$.

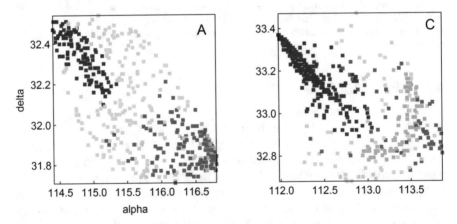

**Fig. 4.4** Model radiant structure for the complete activity period for the model stream $m_4$ in sections *A* and *C* (Fig. 4.2). Cells are colour-coded as a function of the true anomaly of ejection point $\upsilon$ on the cometary orbit: dark blue: $180° < \upsilon < 270°$, i.e., the nucleus is moving away from the aphelion; light blue: $270° < \upsilon < 360°$, i.e., the nucleus is approaching the perihelion; red: $0° < \upsilon < 90°$, i.e., the nucleus passed perihelion; pink: $90° < \upsilon < 180°$, the nucleus is moving to aphelion. For the plot 500 random model meteoroids were taken from the complete sample. Axes labels should not be compared to observations, because the model stream is not calibrated. Modified after Ryabova (2006, fig. 11)

# References

Abedin, A., Spurný, P., Wiegert, P., Pokorný, P., Borovička, J., Brown, P.: On the age and formation mechanism of the core of the Quadrantid meteoroid stream. Icarus **261**, 100–117 (2015). https://doi.org/10.1016/j.icarus.2015.08.016

Hajduková, J.M., Koten, P., Kornoš, L., Tóth, J.: Meteoroid orbits from video meteors. The case of the Geminid stream. Planet. Space Sci. **143**, 89–98 (2017). https://doi.org/10.1016/j.pss.2017.01.004

Ryabova, G.O.: Meteoroid streams: mathematical modeling and observations. In: Verbeeck, C., Wislez, J.M. (eds.) Proceedings of the Radio Meteor School, pp. 67–76 (2006)

Ryabova, G.O.: Mathematical modelling of the Geminid meteoroid stream. Mon. Not. R. Astron. Soc. **375**, 1371–1380 (2007). https://doi.org/10.1111/j.1365-2966.2007.11392.x

Ryabova, G.O.: Model radiants of the Geminid meteor shower. Earth Moon Planets **102**, 95–102 (2008). https://doi.org/10.1007/s11038-007-9180-4

Ryabova, G.O.: A preliminary numerical model of the Geminid meteoroid stream. Mon. Not. R. Astron. Soc. **456**, 78–84 (2016). https://doi.org/10.1093/mnras/stv2626

# Chapter 5
# Visualization of the Results

In this chapter we will consider simple means that do not require dedicated software as, say, (Clark and Wiegert 2014). I use Golden Software Grapher 7.0 (an old version, but it is more than enough for my purposes), but some other scientific graphics software should cope with these 2D graphs.

It is a good practice to begin with a simple scatter plot, where positions of the model meteoroids are plotted in projection to ecliptic plane. This plot enables particles' density variations *along* the orbit to be displayed. It is also convenient to present filaments or particles' concentrations in a stream. Figure 5.1 shows such plot for a Quadrantid model stream. Williams and Ryabova (2011) used a rather artificial model to demonstrate that any initial structure generated by the ejection process will soon be lost. A distinct curved filament seen in the perihelion part of the stream is a slight remnant of this initial structure.

The next step is usually to plot the stream cross-section in the ecliptic plane.

## 5.1 Cross-Sections

### 5.1.1 In the Ecliptic Plane

A stream intersects the ecliptic plane twice: at the ascending and descending nodes. The true anomaly of the descending node is given by

$$\upsilon_n = 3\pi - \omega, \tag{5.1}$$

and for the ascending node

$$\upsilon_n = 2\pi - \omega. \tag{5.2}$$

© The Author(s), under exclusive license to Springer Nature Switzerland AG 2020
G. O. Ryabova, *Mathematical Modelling of Meteoroid Streams*, SpringerBriefs in Astronomy, https://doi.org/10.1007/978-3-030-51510-2_5

**Fig. 5.1** Model Quadrantid stream meteoroids in projection to the ecliptic plane in 2011. 8000 test particles were ejected in 1493 at perihelion of the orbit of (196256) 2003EH1, all with the same speed given by Whipple's formula. All details of the model used see in the work Williams and Ryabova (2011)

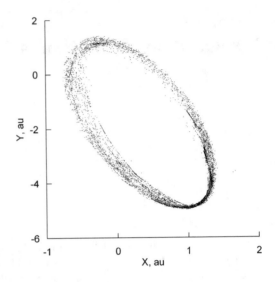

The rectangular coordinates of a node point can be found either using the procedure `aei_xyz` (Appendix A) or by

$$r_n = \frac{a(1 - e^2)}{1 + e \cos \upsilon_n},$$

$$x_n = -r_n \cos \Omega,$$

$$y_n = -r_n \sin \Omega.$$

$$(5.3)$$

The second solution is preferable when you are concerned with speed or precision of your calculations, otherwise use `aei_xyz` (or a similar routine). It is easy to see that the solar longitude $\lambda$ of this point may be expressed by

$$\lambda = \arctan\left(\frac{y_n}{x_n}\right).$$

This type of graph is especially useful when we want to figure out if a stream will be observed on the Earth. Figure 5.2 shows, for example, that the Geminids were not observed on the Earth one thousand years ago, but now they might be observed also on the Venus and Mars.

Certainly, when we consider, say, Mars-encountering meteoroids, the orbital plane of the planet instead of the ecliptic plane should be taken. A good example of the latter see in Christou and Vaubaillon (2011, fig. 2), where Mars-encountering test particles ejected from comet C/2007 H2 (Skiff) are shown. This cross-section is especially interesting by the numerous concentrations of particles within.

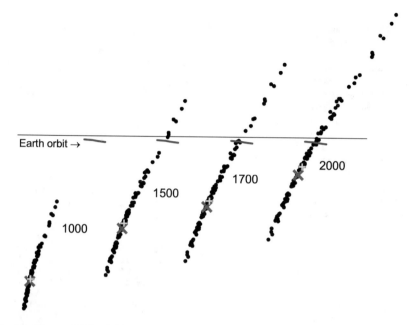

**Fig. 5.2** The model Geminid stream (100 particles, mass = 0.02 g) cross-sections in the ecliptic plane at the descending node in the years 1000, 1500, 1700, and 2000. Nodes of the meteoroids are designated by black dots. The small cyan cross (plus symbol) is the node of Phaethon's orbit, the large red cross is the Geminid's mean orbit node, and the blue line is the Earth's orbit. The horizontal line shows that the cross-sections are aligned. We used the standard heliocentric ecliptic system as reference system. After Ryabova and Rendtel (2018, fig. 2)

## 5.1.2  In a Perpendicular Plane

The cross-section of a stream in a plane perpendicular to the velocity vector of the parent body $V_c$ enables to see the orbits' (not particles') density distribution *across* the stream. All mathematics and computer routines may be found in Appendix D. A similar formulae may be found in the work by Fox et al. (1983). The rectangular coordinate system ($\Xi, E, Z$) is also described in Appendix D. There is no need to go into details now. To understand the figures we need to know only that (1) the orbit's node coordinates are $\xi, \eta$ in the perpendicular plane and (2) the $\xi$ axis is in the plane of the reference orbit, positive towards the anti-Sun direction.

This type of graph is helpful in the analysis of evolution of a stream structure and its evolution. To analyse the density distribution within the cross-section it is convenient to use *equidensities*. The technique of the construction of equidensities is explained by Fig. 5.3. In the beginning *the density matrix* is constructed. A grid is applied over the cross-section, and the number of nodes in every cell (rectangular element of the grid) is calculated. For 5000 test particles model the 50 × 50 matrix

**Fig. 5.3** How to make equidensities. The upper plot shows the model Geminid cross-section in a plane perpendicular to the velocity vector of the reference orbit at the descending node. The stream consists of 1000 orbits of particles with masses $m_3$ ejected at perihelion at $t = 0$ isotropically. The scale is the same in both axes. The cross-section is a base for the density matrix, and hence for the plot of equidensities. More details see in the text

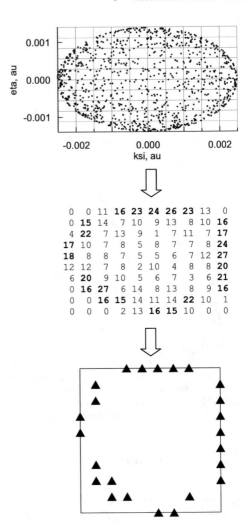

is good enough. Then we 'paint' the matrix with symbols, e.g. by black triangles if the cell has $\geq 15$ nodes as in Fig. 5.3.

An illustrative example of the cross-section evolution with time is furnished by Fig. 5.4. Just two equidensities, designated by black and grey squares, reveal a certain cross-like structure and its transformation with time. The reason of the structure formation we will discuss briefly in Sect. 5.2. Now we are concentrating on the best technique for its visualization. For comparison see Fig. 5.5. It is obvious that it gives us little or nothing to analyse.

*A Technical Note*  To draw the plots in Figs. 5.4, 5.6, and 5.7 I used class scatter plot tool (Golden Software Grapher 7.0). A class scatter plot is a plot with symbols for each XY location based on a required third value (class column).

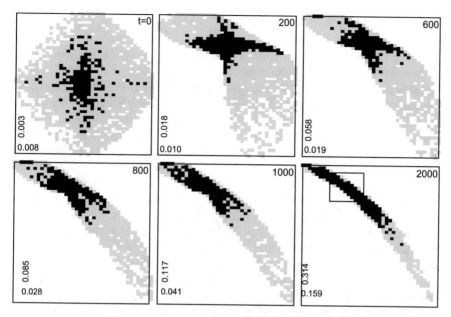

**Fig. 5.4** Cross-sections of the model Geminid stream in a plane perpendicular to the velocity vector of the reference orbit at the descending node. The time in years from the moment of the stream generation is indicated in the top right-hand corner of every panel. The densest part of the stream is shown by black colour. The numbers in the bottom left-hand corners are the length (in au) of the sides of the corresponding rectangle enclosing the cross-section. The mass of particles is $m_3$. Modified after Ryabova (2006, fig. 8)

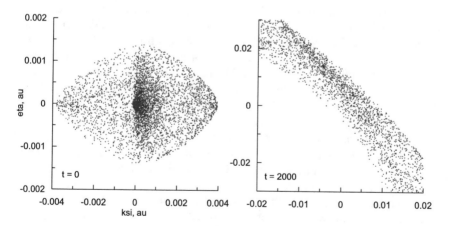

**Fig. 5.5** Cross-sections of the model Geminid stream at the descending node in a plane perpendicular to the reference orbit. The model is the same as in Fig. 5.4 and consists of 5000 orbits. The nodes are shown for $t = 0$ (left plot), and for $t = 2000$ (right plot). For $t = 2000$ only part of the cross-section, marked in Fig. 5.4 by rectangle, is shown

## 5.2   Colour, or A Story About Layers in Geminids

Now, when most of scientific journals are published electronically only, a researcher may present all his/her figures in colour if (s)he wish to do so. Undoubtedly colour assists in explanation of complicated matters to the Reader. However, it is of primary importance that colour aids in revealing of implicit or latent patterns in the data to the Researcher. There are many publications devoted to theory of visualization or data visualization. This short text adds nothing to the theory, but offers an example, when the proper visualization helped to discover a basic feature of a cometary origin stream. And now the real-life story.

The first models of the Geminid meteoroid stream I published about 30 years ago. One of the simulation results was quite unexpected, namely the bimodality of the shower's activity profile. A short publication claiming the fact issued in the journal of International Meteor Society in 1989 after it was declined by a professional journal. Later these results were supported by observations and published in journals and proceedings (e.g., Ryabova 2001a,b). This was the first case in my practice, when something important was discovered first in the model and later in observations.

You may see the bimodal profiles in small panels $A–E$ of Fig. 4.2. Figure 5.4 explains why profiles have two peaks. The mentioned on the page 40 cross-like structure is distinguishable in the cross-section from the very beginning, i.e. at $t = 0$. At $t = 600$ it is 'scissors' rather than 'cross'. In the last panel ($t = 2000$), the fine structure cannot be seen in details because of the scale.

Colour allows to reveal the fine structure better than greyscale. We see that after 2000 years of evolution the stream separates into two layers (Fig. 5.6a, c). But why? Let us address the right panels of Fig. 5.6. They present the same cross-sections as the left panels, but the cells colour code indicates when the particles were ejected: before or after perihelion. The orbital characteristics of the pre- and post-perihelion meteoroids are a bit different. This difference increases with time, resulting in the formations of two layers in the stream. The Earth in its motion crosses these layers one after another (Fig. 4.1), which gives rise to a bimodal activity profile. Note that the cross is weakly pronounced in the case of isotropic meteoroid ejection, whereas it is clearly defined in the case of ejection in the sunward cone.

The story is not finished yet. Before closing the chapter let us turn to Fig. 5.7. It was reasonable to suggest that the pre- and post-perihelion layers have different orbital characteristics. In the model it is really so. The left plot was published in Ryabova (2007). Compare it with the right plot in colour. It is obvious that the patterns in the orbital elements' configurations are more expressive and more understandable. This research problem is still opened. To close it, the similar (or dissimilar) patterns should be revealed in observations.

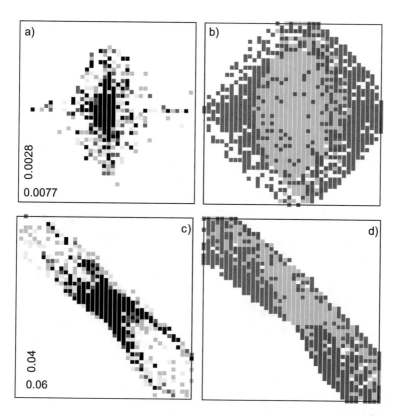

**Fig. 5.6** Left panels: (**a**) and (**c**). Cross-sections of the model Geminid stream at the descending node in a plane perpendicular to the velocity vector of the reference orbit. Three equidensities (yellow, green, red) and the densest part of the stream (black) are shown. (**a**) $t = 0$, i.e., just after formation; (**b**) the stream age is 2000 years; note that only part of the cross-section is shown, i.e., the rectangle on the last panel of Fig. 5.4. Right panels: (**b**) and (**d**). The same sections, but different symbols show the cells (rectangular elements) of the cross-section through which the orbits of particles pass: particles ejected before perihelion passage (blue), after perihelion passage (red), and both before and after the perihelion passage (green). After Ryabova (2006, fig. 9)

## Exercises

5.1 Simulate ejection of 1000 Geminid meteoroids isotropically in perihelion of the parent body. Calculate the ejection speed using the Whipple formula (2.7) assuming that meteoroids are spherical, their density is $1\,\mathrm{g\,cm^{-3}}$, and their mass is $3 \times 10^{-5}\,\mathrm{g}$, and $R_c = 10\,\mathrm{km}$, $\rho_c = 1.0\,\mathrm{g\,cm^{-3}}$. Find the model Geminid cross-section in the ecliptic plane. Repeat the same for the true anomaly of the ejection point $\upsilon = 45°, 90°, 135°, 180°, 225°, 270°$, and $315°$. Make the plots. Explain the results.

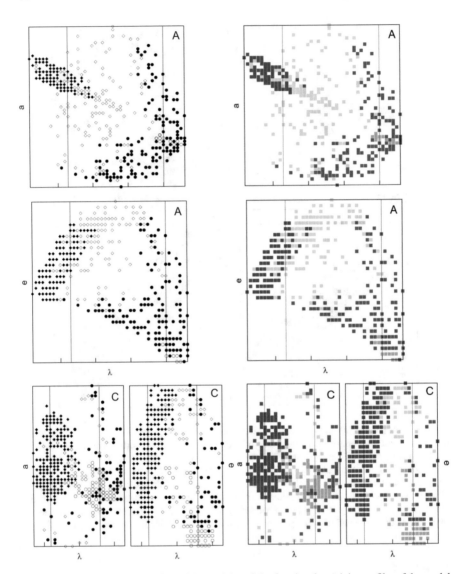

**Fig. 5.7** Orbital elements ($a$ and $e$) of the model particles forming the activity profiles of the model stream $m_4$ in sections A and C (Fig. 4.2) versus the solar longitude. Left black and white plot is after Ryabova (2007, fig. 11). The right plot shows the same, but in colour. Cells are colour-coded as a function of the true anomaly of ejection point $\upsilon$ on the cometary orbit: dark blue: $180° < \upsilon < 270°$, i.e., the nucleus is moving away from the aphelion; light blue: $270° < \upsilon < 360°$, i.e., the nucleus is approaching the perihelion; red: $0° < \upsilon < 90°$, i.e., the nucleus passed perihelion; pink: $90° < \upsilon < 180°$, the nucleus is moving to aphelion. For the plot 500 random model meteoroids were taken from the complete sample. The distance between the ticks in $\lambda$ is equal to $0.5°$. For more details see in the text

5.2 The orbital elements of the asteroid (196256) 2003EH1 are: $a = 3.123$ au, $e = 0.619$, $i = 70.840°$, $\Omega = 282.982°$, $\omega = 171.340$. Simulate ejection of 1000 Quadrantids along the parent body orbit taking that the dust production rate is proportional to $r^{-4}$. Let ejection be isotropic from the sunlit hemisphere in the cone with opening 70°. Calculate the ejection speed using Whipple formula (2.7) assuming that meteoroids are spherical, their density is 1 g cm$^{-3}$, and their mass is $3 \times 10^{-3}$ g, and $R_c = 10$ km, $\rho_c = 1.0$ g cm$^{-3}$. Find sections of the model stream in the ecliptic plane both in the ascending and descending nodes. Designate nodes of meteoroids ejected before perihelion by blue colour, and nodes of meteoroids ejected after perihelion by red colour. Explain the result.

# References

Christou, A.A., Vaubaillon, J.: Numerical modeling of cometary meteoroid streams encountering Mars and Venus. In: Cooke, W.J., Moser, D.E., Hardin, B.F., Janches, D. (eds.) Meteoroids: The Smallest Solar System Bodies. Proceedings of the Meteoroids 2010 Conference, pp. 26–30. National Aeronautics and Space Administration, Huntsville, AL (2011)

Clark, D.L., Wiegert, P.A.: Dynamical modelling of meteoroid streams. In: Jopek, T.J., Rietmeijer, F.J.M., Watanabe, J., Williams, I.P. (eds.) Proceedings of International Conference held at A.M. University in Poznań, August 26–30, 2013, Meteoroids 2013, Wydanictwo Naukowe UAM, Poznań, pp. 275–286 (2014)

Fox, K., Williams, I.P., Hughes, D.W.: The rate profile of the Geminid meteor shower. Mon. Not. R. Astron. Soc. **205**, 1155–1169 (1983). https://doi.org/10.1093/mnras/205.4.1155

Ryabova, G.O.: Mathematical model of the Geminid meteor stream formation. In: Warmbein, B. (ed.) Meteoroids 2001 Conference. ESA Special Publication, vol. 495, pp. 77–81. ESA Publications Division, Noordwijk, The Netherlands (2001a)

Ryabova, G.O.: The Geminid meteor stream activity profile. Solar Syst. Res. **35**, 151–157 (2001b)

Ryabova, G.O.: Meteoroid streams: mathematical modeling and observations. In: Verbeeck, C., Wislez, J.M. (eds.) Proceedings of the Radio Meteor School, pp. 67–76 (2006)

Ryabova, G.O.: Mathematical modelling of the Geminid meteoroid stream. Mon. Not. R. Astron. Soc. **375**, 1371–1380 (2007). https://doi.org/10.1111/j.1365-2966.2007.11392.x

Ryabova, G.O., Rendtel, J.: Increasing Geminid meteor shower activity. Mon. Not. R. Astron. Soc. Lett. **475**(1), L77–L80 (2018). https://doi.org/10.1093/mnrasl/slx205

Williams, I.P., Ryabova, G.O.: Meteor shower features: are they governed by the initial formation process or by subsequent gravitational perturbations? Mon. Not. R. Astron. Soc. **415**(4), 3914–3920 (2011). https://doi.org/10.1111/j.1365-2966.2011.19010.x

# Chapter 6
# Application to Real Streams

According to my personal teaching (and learning) experience, one of the best way to master the streams' modelling is reproducing results of a paper that attracted your particular interest. Since you read as far as the sixth chapter, I hope you are interested. So I suggest the Geminid 2007 model as the concluding exercise to master the studied theory and to test your software.

Below you will find two plans for your modelling practice.

**Reproducing the Geminid 2007 Model**

1. Simulate ejection of 5000 Geminid meteoroids from its parent body using parameters of the Geminid 2007 model (Appendix C).
2. Find the model Geminid cross-section in a plane perpendicular to the reference orbit's velocity vector at the descending node. Plot four equidensities: the densest part of the stream and 3 arbitrary, reproducing Fig. 5.6a. Reproduce Fig. 5.6c.
3. Calculate evolution of the stream using the method of nested polynomials (Sect. 3.3).
4. Make a plot reproducing Fig. 5.6b, d.
5. Calculating the radiant (Sect. 4.3) and reproducing Fig. 4.4 is more difficult task. Do it if you have about 1 month (estimation for students) for this work.
6. Repeat steps 1 and 2 for 10 millions of model meteoroids. Obtain the activity profiles of the model Geminid shower (Sect. 4.1), reproducing panel A of Fig. 4.2.

© The Author(s), under exclusive license to Springer Nature Switzerland AG 2020
G. O. Ryabova, *Mathematical Modelling of Meteoroid Streams*, SpringerBriefs in Astronomy, https://doi.org/10.1007/978-3-030-51510-2_6

**A Model of the Stream of Your Choice**

1. Make a search for publications related to this stream. I recommend NASA's Astrophysics Data System Service. The aim of this search is to construct the physical model (Sect. 1.1.1).
2. Construct an initial model as described in Chap. 2.
3. Calculate the stream evolution to the present epoch as described in Chap. 3.
4. Find the cross-section of the stream in the ecliptic plane. If the Earth passes through the stream, go to the step 5. If not, the most probable solution is to choose another epoch of the stream generation and return to the step 2.
5. Obtain the activity profile of the model meteor shower (Sect. 4.1) and find the radiants (Sect. 4.3). These parameters are the very first to compare with observations.
6. Compare the model activity profile, especially time of the maximum activity with the observed profile. If they are not too different, try to understand how you could fit the model. If they are very different, try to understand why.
7. Compare the model radiant with the observed one.

This is a very approximate plan for a BSc thesis.

# Appendix A
# Transition from the Rectangular Coordinates to the Orbital Elements and Back

In the routines that follow (xyz_aei, aei_xyz) names of input and output values are mostly self-explanatory. Units: length is in [au], speed is in [au d$^{-1}$], angles are in [rad]. Both routines are intended for *elliptical* orbits only.

**Problem** The components of the position and velocity of the model meteoroid in the standard heliocentric ecliptic system are: $x, y, z, V_x, V_y, V_z$. Obtain the Keplerian elements of the meteoroid: $a, e, i, \Omega, \omega, \upsilon$.

The orbital elements of the model meteoroid can be computed from the following set of equations

$$V^2 = V_x^2 + V_y^2 + V_z^2, \tag{A.1}$$

$$a = \frac{r}{(2 - rV^2/\mu)}, \tag{A.2}$$

$$e^2 = 1 - \frac{r^2 V^2 - (\mathbf{r} \cdot \mathbf{V})^2}{\mu a}, \tag{A.3}$$

$$(\mathbf{r} \cdot \mathbf{V}) = x V_x + y V_y + z V_z, \tag{A.4}$$

$$p = a(1 - e^2), \tag{A.5}$$

$$\sin i = \frac{(y V_z - z V_y)^2 + (x V_z - z V_y)^2}{\mu p}, \quad \cos i = \frac{x V_y - y V_x}{\sqrt{\mu p}}, \tag{A.6}$$

$$\sin \Omega = \frac{y V_z - z V_y}{\sqrt{\mu p} \sin i}, \quad \cos \Omega = \frac{x V_z - z V_x}{\sqrt{\mu p} \sin i}, \tag{A.7}$$

$$\sin \upsilon = \frac{\sqrt{p} (\mathbf{r} \cdot \mathbf{V})}{\sqrt{\mu} r e}, \quad \cos \upsilon = \frac{p - r}{r e}, \tag{A.8}$$

© The Author(s), under exclusive license to Springer Nature Switzerland AG 2020
G. O. Ryabova, *Mathematical Modelling of Meteoroid Streams*, SpringerBriefs in Astronomy, https://doi.org/10.1007/978-3-030-51510-2

$$\sin u = \frac{z}{r \sin i}, \qquad \cos u = \frac{x \cos \Omega + x \sin \Omega}{r}, \tag{A.9}$$

$$\omega = u - v. \tag{A.10}$$

Notice that a procedure computing the orbital elements will be called million times during modelling. Therefore, the number of arithmetic operations must be minimal. The more the operations, the greater is the rounding error. One of the possible implementation is the following Delphi-code.

```
procedure xyz_aei(x,y,z,vx,vy,vz: extended;
                    var a,e,i,om,w,v: extended; mu: extended);
{v is the true anomaly}

var     mu1,r1,r2,r3,rr,r,vv2,vv,p,p1,yz,xz,xy,u: extended;
begin
  r2:=x*x+y*y+z*z;
  r:=sqrt(r2);
  rr:=x*vx+y*vy+z*vz;
  vv2:=vx*vx+vy*vy+vz*vz;
  vv:=sqrt(vv1);
  mu1:=sqrt(mu);
  a:=r/(2-r*vv2/mu);
  e:=1-(r2*vv1-rr*rr)/(a*mu); {e^2}
  p:=a*(1-e);
  p1:=sqrt(p);
  e:=sqrt(e);
  yz:=y*vz-z*vy;
  xz:=x*vz-z*vx;
  xy:=x*vy-y*vx;
  i:=arctan(sqrt(yz*yz+xz*xz)/xy);
  if xy<0 then  i:=pi+i;
  om:=arctan(yz/xz);
  if  xz<0 then  om:=pi+om;
  r1:=p-r;
  v:=arctan(p1*rr/(b1*r1));
  if r1<0 then  v:=pi+v;
  r3:=r2*vz-rr*z;
  u:=arctan(z*p1*b1/r3);
  if  r3<0 then  u:=pi+u;
  w:=u-v;
  if w<0 then  w:=2*pi+w;
end;  (*xyz_aei*)
```

**Problem** Obtain the components of the position and velocity $(x, y, z, V_x, V_y, V_z)$ of the model meteoroid in the standard heliocentric ecliptic system, when the Keplerian elements of the meteoroid $(a, e, i, \Omega, \omega, \upsilon)$ are known.

$$r = \frac{a(1 - e^2)}{1 + e \cos \upsilon}, \tag{A.11}$$

$$u = \upsilon + \omega,$$

$$x = r(\cos u \cos \Omega - \sin u \sin \Omega \cos i),$$
$$y = r(\cos u \sin \Omega - \sin u \cos \Omega \cos i), \tag{A.12}$$
$$z = r(\sin u \sin i),$$

$$V^2 = \mu \left( \frac{2}{r} - \frac{1}{a} \right),$$

$$V_r = \sqrt{\frac{\mu}{p}} \, e \sin \upsilon,$$

$$V_n = \sqrt{\frac{\mu}{p}} \, (1 + e \cos \upsilon),$$

$$V_x = x V_r / r + (-\sin u \cos \Omega - \cos u \sin \Omega \cos i) V_n,$$
$$V_y = y V_r / r + (-\sin u \sin \Omega + \cos u \cos \Omega \cos i) V_n, \tag{A.13}$$
$$V_z = z V_r / r + (\cos u \sin i) V_n.$$

Please note repeating calculations of *sin* and *cos* in the following Delphi (Pascal) code. The execution time of trigonometric operations is 100–200 in units of execution time of the addition operation. These expensive operations should be optimized in the first place. The variables $r$ and $V$ (velocity in the following code, not to be confused with v—the true anomaly) are included in the list of the output parameters to avoid double calculations.

```
procedure   aei_xyz(a,e,i,om,w,v,mu: extended;
                    var x,y,z,vx,vy,vz,r,velocity: extended);
{v is the true anomaly}

var     ci,com,cu,si,som,su,p,p1,r1,vr,vn,vrr,u: extended;
begin
   u:=v+w;
   p:=a*(1-e*e);
```

```
    p1:=sqrt(mu/p);
    r1:=1+e*cos(v);
    r:=p/r1;
    vr:=p1*e*sin(v);
    vn:=p1*r1;
    vrr:=vr/r;
    com:=cos(om);  som:=sin(om);
    cu:=cos(u);    su:=sin(u);
    ci:=cos(i);    si:=sin(i);
    x:=r*(cu*com-su*som*ci);
    y:=r*(cu*som+su*com*ci);
    z:=r*su*si;
    vx:=x*vrr+(-su*com-cu*som*ci)*vn;
    vy:=y*vrr+(-su*som+cu*com*ci)*vn;
    vz:=z*vrr+cu*si*vn;
    velocity:=sqrt(vx*vx+vy*vy+vz*vz);
end;  (*aei_xyz *)
```

A test example for procedures `xyz_aei` and `aei_xyz` follows.

```
Orbital elements
a  = 1.27091455141883801E+0000 au
e  = 8.89804910051753838E-0001
i  = 3.88425210090747523E-0001 rad
om = 4.63037130793217223E+0000 rad
w  = 5.62139328511147116E+0000 rad
v  = 3.31817413110187456E+0000 rad

Rectangular coordinates
x  = 1.07262058547459554E+0000 au
y  = 1.80574626047247301E+0000 au
z  = 3.76919057022019077E-0001 au
vx =-5.85266600136740918E-0003 au/day
vy =-2.21696374565765387E-0003 au/day
vz =-2.31262481273512208E-0003 au/day
```

# Appendix B
# Random Number Generation

Any compiler, used for mathematical calculations, provides at least one random number generator, namely the one generating the uniformly distributed real numbers between 0 and 1. In Turbo Delphi, for example, a function `Random` exists, which generates random real $U \in [0, 1)$. So to generate a random angle $\Phi \in [0, 2\pi)$ we just use $\Phi = 2\pi \times U$. Delphi also has a function `RandG(Mean, StdDev)`, producing random numbers with Gaussian distribution with the standard parameters *Mean* and *StdDev* (mathematical expectation and standard deviation). However, sometimes (e.g., in the cases described in Chap. 2) we need another distribution.

Let $x$ be a continuous random variable, $f(x)$ be its probability density function[1], and $F(x)$ be its distribution function.[2] The inverse distribution function $G(\gamma) = x$ can be obtained from the equation $\gamma = F(x)$. Then, if $U$ is a random variable distributed uniformly in $[0, 1)$, you may generate your random variable $x$, using the formula $x = G(U)$. This is an intentionally simplified description. The strict formulation could be found in Forbes et al. (2011).

In the further calculations we will use two basic relations for probability distributions, namely

$$\int_{-\infty}^{+\infty} f(x)dx = 1,$$

(B.1)

and

$$F(x) = \int_{-\infty}^{y} f(y)dy.$$

(B.2)

Now we will consider three examples.

---

[1] Also called probability function.

[2] Also called cumulative distribution function.

© The Author(s), under exclusive license to Springer Nature Switzerland AG 2020
G. O. Ryabova, *Mathematical Modelling of Meteoroid Streams*, SpringerBriefs in Astronomy, https://doi.org/10.1007/978-3-030-51510-2

**Example 1**
**Problem** Generate random angles $T \in [0, \alpha]$, where $\alpha$ is the half-opening of the ejection cone, and $f_T(T) \propto \sin(T)$.

We may write $f(t) = A \sin T$, where $A$ is a constant coefficient. Writing (B.1) for our context

$$\int_0^\alpha A \sin T dt = 1, \tag{B.3}$$

we obtain

$$f(T) = \frac{\sin T}{1 - \cos \alpha}. \tag{B.4}$$

By Eqs. (B.2) and (B.4)

$$F(T) = \int_0^T f(x)dx = \frac{\cos T - 1}{\cos \alpha - 1}. \tag{B.5}$$

Then equating $F(T) = U$ we obtain

$$\cos T = 1 - (1 - \cos \alpha)U. \tag{B.6}$$

It is obvious that $\alpha = \pi$ means isotropic ejection, and for this case $\cos T = 1 - 2U$.

**Example 2**
**Problem** Generate random angles $T \in [0, \alpha]$, where $\alpha$ is the half-opening of the ejection cone, $\alpha < \pi/2$, and $f_T(T) \propto \cos(T) \sin(T)$. In comparison with Example 1, the ejection in the cone is not isotropic, but obeys the cosine law.

Following exactly the technique described in Example 1, we obtain

$$\int_0^\alpha B \cos T \sin T dt = 1.$$

Hence

$$f(T) = \frac{2}{\sin \alpha^2} \sin T \cos T, \qquad F(T) = \frac{\sin T^2}{\sin \alpha^2}.$$

(continued)

In the result

$$\sin T = \sin \alpha \sqrt{U}. \tag{B.7}$$

Since $T < \pi/2$ for the sunlit hemisphere, $\cos T > 0$.

**Example 3**

**Problem** For a given orbit generate the true anomaly $v$ so that the probability density distribution $f_r(r)$ was proportional to $r^{-\delta}$.

Again following exactly the technique described in Example 1, and beginning from $f_r(r) = C \times r^{-\delta}$ we obtain

$$F_r(r) = \frac{r^{1-\delta}}{Q^{1-\delta} - q^{1-\delta}} - \frac{q^{1-\delta}}{Q^{1-\delta} - q^{1-\delta}}, \tag{B.8}$$

where $q$ and $Q$ are the perihelion and the aphelion distances, respectively. Note that the second term in (B.8) is a constant. Then

$$r = \left[ \frac{1}{(Q^{1-\delta} - q^{1-\delta})U + q^{1-\delta}} \right]^{-\frac{1}{\delta-1}}. \tag{B.9}$$

From (A.11) it is easy to find $\cos v$. Since `arccos` returns $v \in [0, \pi]$ in Delphi, we should attribute the minus to the calculated $v$ randomly.

More useful information about Monte Carlo modelling you will find in the books by Skiena (2008) and Kahaner et al. (1989). It is essential also that your read a manual (or Help) on *your* pseudorandom generator for recommendations related to the seed choice and the cycle length.

# Appendix C
# Geminid 2007 Model

The Appendix contains all necessary data to reproduce the Geminid stream model described in Ryabova (2007), including the coefficients for Eq. (3.4). The coefficients are given with 10 significant digits, and it is more than enough for educational purposes. Time in (3.4) is counted in hundreds of years from the starting epoch, so $t$ varies from 0 to 20. Now the Reader can reproduce all results of Ryabova (2007) and/or construct his/her own Geminid model. Before doing this I recommend reading this paper carefully. The later model (Ryabova 2016) used essentially the same parameters.

## C.1 Parameters of the Model 2007

The orbit of Phaethon (Table C.1) calculated for epoch JD 172 1206.3 has been used as the reference orbit.

The ratios of the particle's cross-section to its mass $A/m$ were taken to be 9.375 and 20.198. For spherical uniform particles with a density of $1\,\mathrm{g\,cm^{-3}}$, these values correspond to the masses $m_3 = 2.14 \times 10^{-3}$ g and $m_4 = 2.14 \times 10^{-4}$ g. In the model Ryabova (2016) the 'visual' mass $m_2 = 2.0 \times 10^{-2}$ g ($A/m = 4.45$) was appended. The dust production rate was taken proportional to $r^{-4}$. The ejection speed was calculated using the Whipple formula (2.7), where $R_c = 10\,\mathrm{km}$ and $\rho_c = 1.0\,\mathrm{g\,cm^{-3}}$.

© The Author(s), under exclusive license to Springer Nature Switzerland AG 2020
G. O. Ryabova, *Mathematical Modelling of Meteoroid Streams*, SpringerBriefs
in Astronomy, https://doi.org/10.1007/978-3-030-51510-2

**Table C.1**  Orbital elements
of (3200)Phaethon

$$a = 1.2738933950 \, \text{au}$$
$$e = 0.9008617137$$
$$i° = 14.1805704482$$
$$\Omega° = 326.1214481852$$
$$\omega° = 262.4315315621$$
$$M° = 0.0000134358$$
$$\text{Epoch} = \text{JD } 1721206.3$$

## C.2   Tables of Polynomial Coefficients

See Tables C.2, C.3, C.4, and C.5.

**Table C.2**  Coefficients of nested polynomials (au, deg), $m = 2.14 \times 10^{-3}$ g, $a_i < 1.3344$ au

| $k$ | 0 | 1 | 2 | 3 |
|---|---|---|---|---|
| $a_{1k}$ | −1.315362537E−3 | 1.941945184E−4 | −1.242962326E−3 | 4.146568660E−4 |
| $a_{2k}$ | 2.040874930E−4 | −5.140250619E−4 | 4.289499651E−4 | −1.184492953E−4 |
| $a_{3k}$ | −9.796169962E−7 | 2.756986804E−6 | −2.779502595E−6 | 1.061708945E−6 |
| $e_{1k}$ | −1.939805732E−2 | 4.917517862E−2 | −4.190991670E−2 | 1.192869980E−2 |
| $e_{2k}$ | 1.147024045E−4 | −3.454097334E−4 | 3.584570602E−4 | −1.374416318E−4 |
| $e_{3k}$ | −7.222371863E−7 | 2.649662835E−6 | −3.180258910E−6 | 1.297295238E−6 |
| $i_{1k}$ | 1.384396919E+1 | −3.539969671E+1 | 2.995517680E+1 | −8.444055658E+0 |
| $i_{2k}$ | −9.954728313E−3 | 8.149273451E−2 | −1.556144300E−1 | 9.266555322E−2 |
| $i_{3k}$ | −2.159314180E−3 | 2.348865541E−3 | 1.903874637E−3 | −2.182489177E−3 |
| $\Omega_{1k}$ | −9.227824410E+0 | 1.963028535E+1 | −1.295026611E+1 | 5.703833370E−1 |
| $\Omega_{2k}$ | 1.403570437E+0 | −2.841096287E+0 | 1.594765789E+0 | −1.515214322E−1 |
| $\Omega_{3k}$ | 7.187143888E−3 | −4.074655193E−2 | 5.640574264E−2 | −2.247520916E−2 |
| $\omega_{1k}$ | 8.844550887E+0 | −1.854755122E+1 | 1.203581774E+1 | −3.445876319E−1 |
| $\omega_{2k}$ | −1.386876019E+0 | 2.799270371E+0 | −1.560444401E+0 | 1.424073701E−1 |
| $\omega_{3k}$ | −7.402223375E−3 | 4.137579235E−2 | −5.703968852E−2 | 2.269939972E−2 |

**Table C.3**  Coefficients of nested polynomials (au, deg), $m = 2.14 \times 10^{-3}$ g, $a_i > 1.3344$ au

| $k$ | 0 | 1 | 2 | 3 |
|---|---|---|---|---|
| $a_{1k}$ | $-3.653316733\text{E}-3$ | $4.386710420\text{E}-3$ | $-3.563585239\text{E}-3$ | $7.815213955\text{E}-4$ |
| $a_{2k}$ | $-1.629023893\text{E}-4$ | $3.853163531\text{E}-4$ | $-3.074773647\text{E}-4$ | $8.284950960\text{E}-5$ |
| $a_{3k}$ | $6.834727120\text{E}-6$ | $-1.486776336\text{E}-5$ | $1.049625840\text{E}-5$ | $-2.276036614\text{E}-6$ |
| $e_{1k}$ | $-6.285011903\text{E}-3$ | $1.378383711\text{E}-2$ | $-1.040243551\text{E}-2$ | $2.664038015\text{E}-3$ |
| $e_{2k}$ | $7.327311057\text{E}-4$ | $-1.757683815\text{E}-3$ | $1.437622915\text{E}-3$ | $-4.132847941\text{E}-4$ |
| $e_{3k}$ | $-1.909648988\text{E}-5$ | $4.448848155\text{E}-5$ | $-3.504280212\text{E}-5$ | $9.413124271\text{E}-6$ |
| $i_{1k}$ | $1.347247034\text{E}+0$ | $-2.815764158\text{E}+0$ | $1.805024575\text{E}+0$ | $-3.878432641\text{E}-1$ |
| $i_{2k}$ | $4.865435571\text{E}-2$ | $-5.187765383\text{E}-2$ | $-5.666347964\text{E}-2$ | $6.905170496\text{E}-2$ |
| $i_{3k}$ | $2.643127096\text{E}-3$ | $-9.938210031\text{E}-3$ | $1.238715683\text{E}-2$ | $-5.162341913\text{E}-3$ |
| $\Omega_{1k}$ | $-5.208351595\text{E}+0$ | $7.336828325\text{E}+0$ | $-1.230580679\text{E}+0$ | $-2.986103664\text{E}+0$ |
| $\Omega_{2k}$ | $8.476141891\text{E}-1$ | $-1.002666036\text{E}+0$ | $-2.534243614\text{E}-1$ | $4.325130123\text{E}-1$ |
| $\Omega_{3k}$ | $-1.399807823\text{E}-2$ | $-1.826921279\text{E}-3$ | $3.367498604\text{E}-2$ | $-1.828657804\text{E}-2$ |
| $\omega_{1k}$ | $5.514274383\text{E}+0$ | $-8.105256905\text{E}+0$ | $1.967353400\text{E}+0$ | $2.722149367\text{E}+0$ |
| $\omega_{2k}$ | $-8.724643279\text{E}-1$ | $1.061159254\text{E}+0$ | $2.067025760\text{E}-1$ | $-4.197411006\text{E}-1$ |
| $\omega_{3k}$ | $1.362152110\text{E}-2$ | $2.853172187\text{E}-3$ | $-3.463480813\text{E}-2$ | $1.860023098\text{E}-2$ |

**Table C.4**  Coefficients of nested polynomials (au, deg), $m = 2.14 \times 10^{-4}$ g, $a_i < 1.4687$ au

| $k$ | 0 | 1 | 2 | 3 |
|---|---|---|---|---|
| $a_{1k}$ | $-1.152788399-3$ | $-3.392465556\text{E}-3$ | $6.967041792\text{E}-4$ | $-5.100999929\text{E}-5$ |
| $a_{2k}$ | $1.189708402-4$ | $-3.250006374\text{E}-4$ | $3.041314405\text{E}-4$ | $-9.607206363\text{E}-5$ |
| $a_{3k}$ | $-5.164725875-6$ | $1.605230852\text{E}-5$ | $-1.704506555\text{E}-5$ | $6.314766440\text{E}-6$ |
| $a_{4k}$ | $9.580727025-8$ | $-2.965219947\text{E}-7$ | $3.101962646\text{E}-7$ | $-1.107713324\text{E}-7$ |
| $e_{1k}$ | $-5.802854015-3$ | $1.322252603\text{E}-2$ | $-1.099598309\text{E}-2$ | $3.155875862\text{E}-3$ |
| $e_{2k}$ | $8.262029491-5$ | $-2.703141860\text{E}-4$ | $2.972849549\text{E}-4$ | $-1.204019718\text{E}-4$ |
| $e_{3k}$ | $-9.297855036-7$ | $3.010529695\text{E}-6$ | $-3.350294639\text{E}-6$ | $1.342278973\text{E}-6$ |
| $e_{4k}$ | $-8.984923845-9$ | $2.613196316\text{E}-8$ | $-2.545793526\text{E}-8$ | $8.226009597\text{E}-9$ |
| $i_{1k}$ | $3.491541778+0$ | $-9.681839407\text{E}+0$ | $8.854329248\text{E}+0$ | $-2.721891374\text{E}+0$ |
| $i_{2k}$ | $-2.328106015-1$ | $7.356690673\text{E}-1$ | $-7.926318882\text{E}-1$ | $2.980921802\text{E}-1$ |
| $i_{3k}$ | $1.416514092-2$ | $-4.439672020\text{E}-2$ | $4.666558683\text{E}-2$ | $-1.651952709\text{E}-2$ |
| $i_{4k}$ | $-2.728120516-4$ | $8.303767357\text{E}-4$ | $-8.412110503\text{E}-4$ | $2.832581853\text{E}-4$ |
| $\Omega_{1k}$ | $1.877628311+0$ | $-6.846421278\text{E}+0$ | $7.300501556\text{E}+0$ | $-4.274486024\text{E}+0$ |
| $\Omega_{2k}$ | $-2.127204742-1$ | $7.886762824\text{E}-1$ | $-9.176747936\text{E}-1$ | $3.429258910\text{E}-1$ |
| $\Omega_{3k}$ | $1.245273191-2$ | $-2.715711110\text{E}-2$ | $1.257978442\text{E}-2$ | $3.020038774\text{E}-3$ |
| $\Omega_{4k}$ | $1.203576737-4$ | $-7.382282999\text{E}-4$ | $1.207847140\text{E}-3$ | $-6.043152568\text{E}-4$ |
| $\omega_{1k}$ | $-1.789856025+0$ | $6.647985376\text{E}+0$ | $-7.056966600\text{E}+0$ | $4.152293259\text{E}+0$ |
| $\omega_{2k}$ | $2.165292315-1$ | $-7.990076969\text{E}-1$ | $9.267364986\text{E}-1$ | $-3.453397853\text{E}-1$ |
| $\omega_{3k}$ | $-1.277723547-2$ | $2.815174820\text{E}-2$ | $-1.361920999\text{E}-2$ | $-2.646467303\text{E}-3$ |
| $\omega_{4k}$ | $-1.155420885-4$ | $7.235313839\text{E}-4$ | $-1.192604521\text{E}-3$ | $5.989216936\text{E}-4$ |

**Table C.5** Coefficients of nested polynomials (au, deg), $m = 2.14 \times 10^{-4}$ g, $a_i > 1.4687$ au

| $k$ | 0 | 1 | 2 | 3 |
|---|---|---|---|---|
| $a_{1k}$ | 2.871087426−3 | −1.177981223E−2 | 6.542025104E−3 | −1.412975880E−3 |
| $a_{2k}$ | −9.364415268−4 | 2.002277163E−3 | −1.401939504E−3 | 3.196315820E−4 |
| $a_{3k}$ | 7.850805120−5 | −1.431514524E−4 | 8.243660034E−5 | −1.398141885E−5 |
| $a_{4k}$ | −8.300769022−7 | 1.113151048E−6 | −2.944085034E−7 | −6.154569216E−8 |
| $e_{1k}$ | −1.495906763−2 | 2.578540830E−2 | −1.492978120E−2 | 2.888931154E−3 |
| $e_{2k}$ | 2.123897097−3 | −4.408672315E−3 | 3.110580569E−3 | −7.622791600E−4 |
| $e_{3k}$ | −5.517760273−5 | 1.151887213E−4 | −8.102907901E−5 | 1.936030944E−5 |
| $e_{4k}$ | −6.049403833−7 | 1.171652590E−6 | −7.605362592E−7 | 1.658700871E−7 |
| $i_{1k}$ | 1.042721930+1 | −1.432816324E+1 | 5.353330987E+0 | −3.725579908E−1 |
| $i_{2k}$ | 5.393635131−1 | −1.219712204E+0 | 7.308476865E−1 | −7.484213975E−2 |
| $i_{3k}$ | −4.173145818−2 | 5.604325614E−2 | −7.095803259E−3 | −8.956539456E−3 |
| $i_{4k}$ | 4.794307826−4 | −3.614052196E−5 | −8.228159313E−4 | 4.377034054E−4 |
| $\Omega_{1k}$ | −4.193513567+1 | 7.268037734E+1 | −3.856963197E+1 | 3.899661332E+0 |
| $\Omega_{2k}$ | 1.512836912+1 | −2.359384831E+1 | 1.065627149E+1 | −1.071447228E+0 |
| $\Omega_{3k}$ | −9.196386305−1 | 1.307244470E+0 | −4.943998026E−1 | 2.345156057E−2 |
| $\Omega_{4k}$ | 1.677480625−2 | −2.144946255E−2 | 6.038869303E−3 | 4.588193251E−4 |
| $\omega_{1k}$ | 4.360714736+1 | −7.604606650E+1 | 4.094271561E+1 | −4.504460402E+0 |
| $\omega_{2k}$ | −1.518897877+1 | 2.372742911E+1 | −1.075382893E+1 | 1.095241436E+0 |
| $\omega_{3k}$ | 9.154336299−1 | −1.298067053E+0 | 4.875273609E−1 | −2.167181852E−2 |
| $\omega_{4k}$ | −1.662089452−2 | 2.112244203E−2 | −5.803766402E−3 | −5.162455258E−4 |

# Appendix D
# Cross-Section in a Perpendicular Plane

**Problem** Let the reference orbit be given by its velocity $\mathbf{V_c}$ and position $\mathbf{r_c}$ vectors. Let the orbit of a model meteoroid have osculating orbital elements $a, e, i, \Omega, \omega$. Both the vectors and elements are given in standard heliocentric ecliptic system. Find the coordinates $(\xi, \eta, \zeta)$ of the orbit's node with the plane perpendicular to $\mathbf{V_c}$ at a position $\mathbf{r_c}$. The origin of the rectangular coordinate system $(\Xi, E, Z)$ is at the non-Sun end of $\mathbf{r_c}$. The $\xi$ axis is in the plane of the reference orbit, positive towards the anti-Sun direction. The $\zeta$ axis is along $-\mathbf{V_c}$, and all three axes form a right-handed rectangular coordinate system.

This problem might be splitted into two independent sub-problems.

**Sub-Problem 1** Let the reference orbit be given by its velocity $\mathbf{V_c}$ and position $\mathbf{r_c}$ vectors. Let the orbit of a model meteoroid have osculating orbital elements $a, e, i, \Omega, \omega$. Find the true anomaly $\upsilon$ of the orbit's node(s) with the plane perpendicular to $\mathbf{V_c}$ at a position $\mathbf{r_c}$. Find coordinates $(x, y, z)$ of this point.

Mathematically we have the intersection of an ellipse with a plane here. In the point-direction form (Korn and Korn 1968, Sec. 3.2-1b) the equation of our perpendicular plane we may write

$$V_{cx}(x - x_c) + V_{cy}(y - y_c) + V_{cz}(z - z_c) = 0. \tag{D.1}$$

Let us rewrite Eq. (D.1) in the more convenient general form

$$A_1 x + B_1 y + C_1 z + D_1 = 0. \tag{D.2}$$

© The Author(s), under exclusive license to Springer Nature Switzerland AG 2020
G. O. Ryabova, *Mathematical Modelling of Meteoroid Streams*, SpringerBriefs in Astronomy, https://doi.org/10.1007/978-3-030-51510-2

The parametric equation of the ellipse ($u$ is the parameter) are

$$x = r(\cos u \cos \Omega - \sin u \sin \Omega \cos i),$$
$$y = r(\cos u \sin \Omega - \sin u \cos \Omega \cos i), \qquad \text{(D.3)}$$
$$z = r(\sin u \sin i).$$

where

$$r = p/(1 + e \cos (u - \omega)),$$
$$u = \upsilon + \omega. \qquad \text{(D.4)}$$

Now we rewrite Eq. (D.3) in another form

$$x = \frac{b_1 \cos u - b_3 \sin u}{1 + b_6 \cos u + b_7 \sin u},$$
$$y = \frac{b_2 \cos u - b_4 \sin u}{1 + b_6 \cos u + b_7 \sin u}, \qquad \text{(D.5)}$$
$$z = \frac{b_5 \sin u}{1 + b_6 \cos u + b_7 \sin u},$$

where

$$b_1 = a(1 - e^2) \cos \Omega,$$
$$b_2 = a(1 - e^2) \sin \Omega,$$
$$b_3 = a(1 - e^2) \sin \Omega \cos i,$$
$$b_4 = a(1 - e^2) \cos \Omega \cos i,$$
$$b_5 = a(1 - e^2) \sin i,$$
$$b_6 = e \cos \omega,$$
$$b_7 = e \sin \omega.$$

Substituting $x$, $y$, and $z$ from (D.5) into (D.2) we obtain

$$E \cos u + F \sin u + D_1 = 0,$$

where

$$E = A_1 b_1 + B_1 b_2 + D_1 b_6,$$
$$F = B_1 b_4 - A_1 b_3 + C_1 b_5 + D_1 b_7.$$

Since

$$\sin 2u = \frac{2\tan u}{1 + \tan^2 u}, \qquad \cos 2u = \frac{1 - \tan^2 u}{1 + \tan^2 u},$$

with a little manipulation we obtain

$$\tan \frac{u}{2} = \frac{-F \pm \sqrt{F^2 - D_1^2 + E^2}}{D_1 - E}. \tag{D.6}$$

The desired true anomaly $v$ we find from Eq. (D.4). The routine that implements the above calculations is

```
procedure cross1(x,y,z,vx,vy,vz,a,e,i,om,w: extended;
                 var v:extended);
(*Input:
  Plane: specified by the point (x,y,z) and by the vector
         (Vx,Vy,Vz).
  Orbit: Kepler's orbital elements a,e,i,Om,w
Output:
  v - true anomaly of the point on the orbit, where it
      intersects the plane (two points in reality: v and -v).
Units: length in [au], speed in [au/day], angles in [rad] *)


var   b0,b1,b2,b3,b4,b5,b6,b7,b8,b9,b10: extended;
begin
  b0:=a*(1-e*e);
  b1:=b0*cos(om);
  b2:=b0*sin(om);
  b3:=b2*cos(i);
  b4:=b1*cos(i);
  bBM:=b0*sin(i);
  b6:=e*cos(w);
  b7:=e*sin(w);
  b8:=-(vx*x+vy*y+vz*z);           {D}
  b9:=vx*b1+vy*b2+b8*b6;           {E}
  b10:=vy*b4-vx*b3+vz*b5+b8*b7;    {F}
  v:=2*ArcTan2( (-b10+sqrt(sqr(b10)-sqr(b8)+sqr(b9)))/
                (b8-b9) )-w;
end; {cross1}
```

The routine `cross1` calculates the true anomaly $v$. Coordinates $(x, y, z)$ of this point are calculated with the routine `aei_xyz` (Appendix A).

**Sub-Problem 2**  Transformation $(x, y, z) \Rightarrow (\xi, \eta, \zeta)$.

To make this transformation we have to find directional cosines for the $\xi$ axis, $\eta$ axis, and $\zeta$ axis.

*Axis* $\xi$. The plane of the reference orbit may be specified by two points in this plane, namely $P_1(0, 0, 0)$ and $P_2(x_c, y_c, z_c)$, and the vector $\mathbf{V_c}$ (Bronshtein et al. 2007, eq. 3.376a). So the equation of the plane is

$$\begin{vmatrix} x & y & z \\ x_c & y_c & z_c \\ V_{cx} & V_{cy} & V_{cz} \end{vmatrix} = 0.$$

And in the general form

$$A_2 x + B_2 y + C_2 z = 0, \tag{D.7}$$

where

$$A_2 = y_c V_{cz} - z_c V_{cy}, \quad B_2 = z_c V_{cx} - x_c V_{cz}, \quad C_2 = x_c V_{cy} - y_c V_{cx}.$$

The $\xi$ axis is the intersection of two planes (Korn and Korn 1968, Sec. 3.3-1a) given by the set of equations (D.2) and (D.7). This line also may be given by its direction vector $\mathbf{K}(K_x, K_y, K_z)$ (Korn and Korn 1968, Sec. 3.3-1c), where

$$K_x = B_1 C_2 - C_1 B_2, \quad K_y = C_1 A_2 - A_1 C_2, \quad K_z = A_1 B_2 - B_1 A_2. \tag{D.8}$$

*Axis* $\eta$ The direction vector for this axis is easy to find through the vector product $\mathbf{E} = \mathbf{K} \times \mathbf{V_c}$. Hence the transformation equations are

$$\xi = l_1(x - x_c) + m_1(y - y_c) + n_1(z - z_c),$$
$$\eta = l_2(x - x_c) + m_2(y - y_c) + n_2(z - z_c), \tag{D.9}$$
$$\zeta = l_3(x - x_c) + m_3(y - y_c) + n_3(z - z_c),$$

where the directional cosines for the $\xi$ axis, $\eta$ axis, and $\zeta$ axis are

$$l_1 = \frac{K_x}{|K|}, \quad m_1 = \frac{K_y}{|K|}, \quad n_1 = \frac{K_z}{|K|},$$

$$l_2 = \frac{E_x}{|E|}, \quad m_2 = \frac{E_y}{|E|}, \quad n_2 = \frac{E_z}{|E|},$$

$$l_3 = \frac{-V_{cx}}{|V_c|}, \quad m_3 = \frac{-V_{cy}}{|V_c|}, \quad n_3 = \frac{-V_{cz}}{|V_c|}.$$

And the routine implementing the transformation $(x, y, z) \Rightarrow (\xi, \eta, \zeta)$ is

```
procedure cross2(x0,y0,z0,Vx0,Vy0,Vz0,x,y,z:extended;
              var ksi, eta, dzeta: extended);
{Transformation from (x,y,z) of the standard ecliptic system
  to (ksi,eta,dzeta) of the plane specified by
```

```
the point (x0,y0,z0) and by the vector (Vx0,Vy0,Vz0).
Units: [au] and [au/day].}

var A1,B1,C1,rec_V0,A2,B2,C2,rec_K,Kx,Ky,Kz,rec_E,Ex,Ey,Ez,
    l1,l2,l3,m1,m2,m3,n1,n2,n3: extended;
begin
 {Perpendicular plane}
   A1:=Vx0;
   B1:=Vy0;
   C1:=Vz0;
   rec_V0:=1/sqrt(sqr(Vx0)+sqr(Vy0)+sqr(Vz0));
 {Plane of the reference orbit}
   A2:=y0*Vz0-z0*Vy0;
   B2:=z0*Vx0-x0*Vz0;
   C2:=x0*Vy0-y0*Vx0;
 {KSI axis}
   Kx:=B1*C2-C1*B2;
   Ky:=C1*A2-A1*C2;
   Kz:=A1*B2-B1*A2;
   rec_K:=1/sqrt(sqr(Kx)+sqr(Ky)+sqr(Kz));
 {ETA axis}
   Ex:=Ky*Vz0-Kz*Vy0;
   Ey:=Kz*Vx0-Kx*Vz0;
   Ez:=Kx*Vy0-Ky*Vx0;
   rec_E:=1/sqrt(sqr(Ex)+sqr(Ey)+sqr(Ez));
 {directional cosines}
   l1:=Kx*rec_K;    m1:=Ky*rec_K;    n1:=Kz*rec_K;
   l2:=Ex*rec_E;    m2:=Ey*rec_E;    n2:=Ez*rec_E;
   l3:=-VX0*rec_V0; m3:=-Vy0*rec_V0; n3:=-Vz0*rec_V0;
 {Translation of the origin}
   A2:=x-x0; B2:=y-y0; C2:=z-z0;
   ksi:=   l1*A2+m1*B2+n1*C2;
   eta:=   l2*A2+m2*B2+n2*C2;
   dzeta:=l3*A2+m3*B2+n3*C2;
end; (*cross21*)
```

# References

Bronshtein, I.N., Semendyayev, K.A., Musiol, G., Muhlig, H.: Handbook of mathematics. Springer, Berlin (2007). https://doi.org/10.1007/978-3-540-72122-2

Forbes, C., Evans, M., Hastings, N., Peacock, B.: Statistical distributions. 4th edition. Wiley, New York (2011)

Kahaner, D., Moler, C., Nash, S.: Numerical methods and software. 4th edition. Prentice-Hall, New Jersey (1989)

Korn, G.A., Korn, T.M.: Mathematical Handbook for Scientists and Engineers: Definitions, Theorems, and Formulas for Reference and Review, 2nd enl. and rev. edn. McGraw-Hill, New York (1968)

Ryabova, G.O.: Mathematical modelling of the Geminid meteoroid stream. Mon. Not. R. Astron. Soc. **375**, 1371–1380 (2007). https://doi.org/10.1111/j.1365-2966.2007.11392.x

Ryabova, G.O.: A preliminary numerical model of the Geminid meteoroid stream. Mon. Not. R. Astron. Soc. **456**, 78–84 (2016). https://doi.org/10.1093/mnras/stv2626

Skiena, S.S.: The algorithm design manual. 2nd edition. Springer, London (2008)

# Index

Printed in the United States
By Bookmasters